图书在版编目（CIP）数据

翻新酒店 / 鄢格编译. -- 沈阳 ：辽宁科学技术出版社，2012.1
　ISBN 978-7-5381-6584-5

　Ⅰ . ①翻… Ⅱ . ①鄢… Ⅲ. ①饭店－室内装饰设计－世界 Ⅳ . ①TU247.4

中国版本图书馆CIP数据核字(2011)第194646号

出版发行：辽宁科学技术出版社
　　　　　（地址：沈阳市和平区十一纬路29号　邮编：110003）
印 刷 者：利丰雅高印刷（深圳）有限公司
经 销 者：各地新华书店
幅面尺寸：230mm×290mm
印　　张：34
插　　页：4
字　　数：50千字
印　　数：1～2000
出版时间：2012年 1 月第 1 版
印刷时间：2012年 1 月第 1 次印刷
责任编辑：陈慈良
封面设计：杨春玲
版式设计：杨春玲
责任校对：周　文
书　　号：ISBN 978-7-5381-6584-5
定　　价：268.00元

联系电话：024-23284360
邮购热线：024-23284502
E-mail: lnkjc@126.com
http://www.lnkj.com.cn
本书网址：www.lnkj.cn/uri.sh/6584

OLD HOTEL **NEW FACE**

Editor & Translator: Orange Yan

翻 新 酒 店

鄢格 编/译

辽宁科学技术出版社

Contents 目录

Brief Personal History of an Architectural and Cultural Revaluation in Chiloé
智鲁岛上一次建筑与文化层面上的重塑之旅

I have been asked to write an essay about hotel renovation, yet the text you are about to read is more from an architect's testimony and his relationship with the context in an urban and architectural space built with wood, on stilts, on the shore of an archipelago located in the insular Patagonia in southern Chile.

Palafittes in Castro have been historically support for insular life, and the highest expression of its culture's identity and singularity. In this contemporary architectural revaluation is that inserted an architectural renovation that transforms a precarious pile dwelling into a sophisticated and pertinent Boutique Hotel.

The Enchantment
June 21st, the deep winter begins in the island of Chiloé, though it is no raining, and a bright light filters through the mist illuminating the blue sea, broken by green slopes that contain the Fjord of Castro, where tides rise and fall four times a day.

Perhaps the magic light of this insular Patagonia is the same that lit up the Palafittes in Castro when I saw them for the first time thirty five years ago, while as a slow rain fell I was captivated by the beauty of this precarious architecture on stilts that embraced life over the sea.

It happens that I was born in the North of Chile in the mid twentieth century, and lived much of my childhood and youth in the modern and planned city of El Salvador, to then go and study architecture at the University of Chile in Valparaiso, where I learned to appreciate spontaneous architecture built with wood on the Hills of the main port in the country.

And when I saw them for the first time, I never imagined that Palafittes would constitute an architectural centrepiece of what was going to be our work and our lives in the future.

My interests back then had to do with the "architecture without architects" which was the same type that in this case had given physical existence to such unique neighbourhoods, and had had as essential material the wood coming from the native forests, supported by a wood work experience of more than 500 years, and that we as young architects who came to work in the island of Chiloé, were going to embrace to develop our works.

Palafitte, the Fact
Back in those years, the last fourth of the twentieth century, there were six neighbourhoods of Palafittes built along the street that bordered the city, connecting it with the port and the market. Five of them were pile dwelling neighbourhoods that had become a part of the urban character of the city, plus a neighbourhood of large wooden barracks on stilts where there were furniture factories, sawmills, turneries and storage for wood and potatoes.

In our considerations at Taller Puertazul we came to the conclusion that in conceptual terms the Palafittes as constructions of wood that can adapt to the topography of the ground, are the natural response to inhabit the shore, that space between the highest and the lowest tide defining the spatial dimension and rhythm of the archipelago, located in the inner sea contained between two mountain ranges, the Coast range and the Andes.

我受邀撰写一篇关于酒店改造方面的文章，下面您将读到的内容更多是从建筑师的亲身实践出发，讲述智利南部巴塔哥尼亚群岛"桩住宅"改建成酒店的过程，着重强调建筑与周围环境的关系。

卡斯特罗地区的"桩住宅"（建造在水上或沙滩上由木桩支撑的住宅）是小岛生活的重要组成部分，同时也是该地区的特色文化。我的任务便是将一摇摇欲坠的"桩住宅"改造成精美并与周围环境相宜的精品酒店。

背景——小岛的魅力
6月21日，小岛已进入了深冬时节，明亮的阳光穿过薄雾洒在蓝色的海面上，青山渐渐呈现，每天四次的潮汐更是增添了独特的魅力。

35年前，我第一次来到卡斯特罗看到的是同样的景象，阳光一如往日。当细雨洒落的时候，我彻底沉醉在这一特色建筑所特有的魅力之中，它承载了海滨生活的全部。

20世纪中期，我出生在智利北部的萨尔瓦多市，童年和青年时期全部在现代化城市中度过，之后在位于瓦尔帕莱索市的智利大学学习建筑设计，正是在那里我知道了那些建在港口附近小山上的"桩住宅"。

当我第一次见到这些建筑的时候，从来没有想过有一天它们将会成为我的工作和生活的对象。

我对建筑的兴趣源于《无需建筑师的建筑》一书，正犹如岛上这些独特的住宅一般，其构造材料完全来自于当地森林并采用木桩进行支撑，能够承受住500年时光的雕琢。作为年轻建筑师的我们打算在这里开始打造属于自己的设计。

桩住宅——真实的存在
20世纪后期，桩住宅沿着与城市毗邻的街道而兴起，共形成了6个社区，将港口和市场连结起来。其中5个住宅区已成为城市特色的一部分，另外一个社区则由工厂、锯木厂、车床厂及仓库等构成。

我们从曾在该地区改造过的一个项目中发现，理论上讲，桩住宅这种木质结构很容易适应本地的地势，是海岸建筑结构中最自然的形式。海岸，作为潮汐变化形成的区域决定着空间的体积与岛屿的排列节奏。如此一来，这里位于两大山脉之间的内海海岸则更具优势。

我们工作的出发点是以桩住宅的理念为依托，提出打造公共建筑的必要性与目的性。当然，这无异于制造一种声响，让现代化建筑在这里能够合理的存在并植入其中，但同时更需强调当地木材、传统技术以及木工工艺的运用。

策略——重新诠释
我们致力于使现代风格融入到智鲁岛的建筑形式中，那么对于传统技术进行重新诠释便是最为恰当的策略。经过协商，我们成功说服当地军事当局在卡斯特罗城名为"Dalcahue"的小镇建造一个船员避难所，这将成为岛屿文化中最为传奇的一笔，让船员在恶劣天气之中有一个安全的休息空间。

这一建筑忠实于当地的文化意蕴，因为从伟大摄影师Mr. Gilberto Provoste于20世纪30年代在城市中心的Lillo大街拍摄的照片中便可找到类似的结构。

"桩住宅"社区的发展
19世纪末20世纪初期港口开始发展，为航海社区的形成奠定了基础。这些"桩住宅"建筑多为两层或三层，屹立于海滨上，突出当地的新古典主义风格。建筑功能多种多样，包括仓库、小酒馆、酒店、养老院等，成为港口文化的一部分。

And for this, our first works were marked by the necessity and the intention of making public architecture with the concept of Palafitte, which implied to reinterpret a posture, a way of making and a way of locating, allowing the contemporary work to settle and root in a true manner, as well as the reiteration in the use of native wood, traditional technologies and appreciation for carpentry work.

Reinterpretation as Strategy

This also allowed our contemporary work to become part of the architectural continuum of Chiloé, and the re-interpretation of traditional typologies was the most appropriate way to achieve it.

Through these arguments we managed to convince the military authorities of that time, that in Dalcahue, a town close to the city of Castro, it was necessary to build a shelter for sailors from Chiloé. This was an architectural space that would welcome the deepest and most wonderful of the insular life, a public fogón on stilts that would allow seafarers to shelter from bad weather.

This Palafitte contributed to the cultural dimension of the place, which was also possible to observe in the sepia coloured photographs of Lillo Street in the city of Castro, taken in the 1930s by that the great graphic chronicler, Mr. Gilberto Provoste.

Palafitte Neighbourhoods

The booming port of the city at the end of the nineteenth century and beginning of the twentieth century, gave place to the appearance of sailing neighbourhoods, or houses of the sea, which were large two and three

如此一来，这一建筑样式对于城市居民来说已不再陌生。他们有一天也会将自己的家建在城市的海滩上，正如土生土长的本地人一样，使其成为城市生活的独特体验。

或许是当地马铃薯种植业（当地主要农业经济来源）受到枯萎病的侵蚀，因此，许多家庭移民到城市中。同其他地区那些选择在郊区落户的移民者不同，他们将家安在海滨上。在这里，他们不断复制着自己已有的文化。他们沿街开辟果园，天井被改造成露台，篱笆形成了栅栏，在住宅之间看到猪圈或者家禽饲养场也不必惊奇。这就是我在20世纪70年代看到的"桩住宅"。

"桩住宅"的消失
20世纪70年代末，政府宣布将拆除这些"桩住宅"，我们因此更多地接触到这一建筑样式和周围的居民。事实证明，"桩住宅"是这一以船为主要交通工具的城市中最为自然的建筑形式。

20世纪30年代，火车的兴起大大地改变了街道两侧的建筑结构。火车沿着住宅从街道驶向港口，因此居民不得不选择离开。

此外，1960年的大地震与海啸更是摧毁了城市中的大部分住宅。巨大的海浪席卷，使它们从城市和城镇的边缘消失。梅徐克市和卡斯特罗因位于海岛中心而得以幸存，大部分住宅只是被淹没。

"桩住宅"的维护
在我们看来，将"桩住宅"铲除似乎违背了常理，因此，我们决定维护它们的存在。我们提出将其修缮，让其行使最基本的功能。它们是智鲁岛文化标识以及当地居民（这一建筑传统的继承者）的重要组成部分。

在谈判进行阶段中的一个晚上，火警响彻了整条街道。"桩住宅"被烧掉了，对于这一易燃的结构我们无能为力。

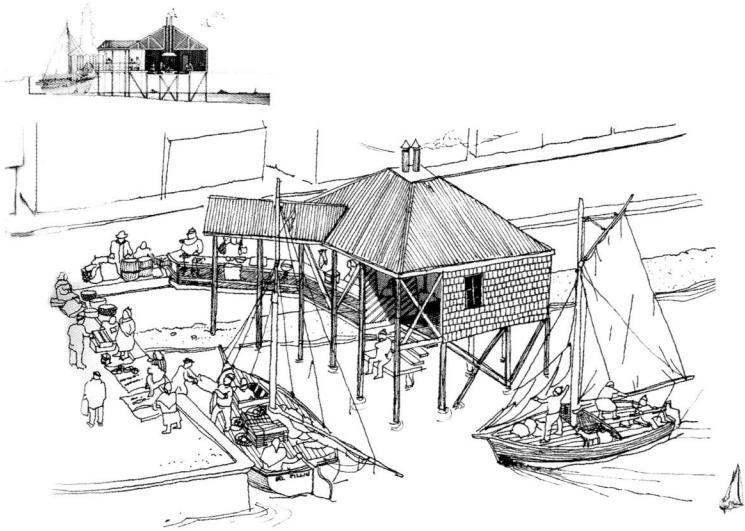

storeys constructions with regional neoclassic aesthetics, built on stilts over the shore, to host potato storage barns, warehouses, taverns, hotels and pensions that were part of the boisterous port world.

Because of this, it was not strange for the inhabitants of a city submerged in its own urban body, that one day any particular family could settle on the shores of the city, just as indigenous people would have done, building their homes on stilts and from their rural conditions, become part of the urban experience.

It was perhaps the blight that destroyed the potato plantations, main source of rural economy, which led country families to migrate to the city. In any other city in the world, they would have located in the suburban areas, but here they did it in the most relevant space in the geography, and from the ductility given by the architecture made of wood; they reproduced their own cultural codes on the shore. Reiterating typological models, along the street they made their orchards, the patio became a terrace and the fence became a railing, and therefore it was not strange to find a pigsty or poultry yard in between the stilts, as I once did back at the end of the 1970s, when Palafittes revealed themselves to us with all their conditions and origins.

Palafitte Eradication

However, it was a municipal decree of eradication of the Palafittes, issued at the end of the 1970s, which led us to get more involved with them and with their people. It turns out that the Palafittes were a natural condition of the city, where the main traffic was by ship and the boat, the natural vehicle as in Venice.

Not even the irruption of the narrow gauge train built in the 1930s had significantly changed the street. All the more, it had consolidated it, because the train followed the street to the port along the decorated façades of those once rural houses. Meanwhile, on the opposite side, the sea continued to be the back yard, with its persistent rural world connected to fishing and shell fishing.

最后，位于Pedro Aguirre Cerda和Lillo大街上的住宅被彻底拆除，居民被转移到其他社区，一部分则选择生活在船上。我们依然工作着，致力于重新运用"桩住宅"。

城市—建筑改造
就在这个时候，我们接到任务设计一家餐厅并已得到海军的许可。我们提出将其打造成"木质的贝壳"。

然而，新的当局却反对重新打造"桩住宅"样式结构，理由是以便于突出海滨的交通，并可以欣赏南部的自然景观。

20世纪90年代，一组深知"桩住宅"这一结构的建筑及文化价值的西班牙建筑师从欧洲经济共同体获得资金对其进行修复，并成功得到了当地政府的支持。

之后，共有两家建筑公司开始工作。我们负责Gamboa社区，提出充分运用回收来的材料，在每间房屋内修建卫生间并重建朝海一面的外观。在工作过程中，经由智利林业所的介绍，我们同瑞典涂料公司合作，打造了一个彩色的"桩住宅"。

传统的价值
必须清楚知道的是："桩住宅"是智鲁岛上最能够体现出木质结构的建筑及文化样式，并处于不断地演进和变化中。这从住宅区的老照片中便能判断出来，其中一些甚至可以称为"古董"。谷草、织物、木材以及刺绣装饰同样成为永恒价值的一部分。当然，除了这些，不同住宅之间的空隙，或是用于储藏木柴或者用于观看景观，更是深化了城市中的地域特色。

拥有传统
在这里，我要讲述一个故事：8年前，一座独立矗立在海滨的古老"桩住宅"即将被拆除，我恰好想买一些废弃的门窗。最后，我同家人决定将整个住宅买下来居住。从此，我彻底明白了修缮一座古老的木建筑需要花费大量的精力和资源。

不过，最让我们着迷的是在涨潮的时候，海水漫到露台，落潮之后会看到鸟类出来觅食和不同的甲壳纲类动物。

面对着天堂一般的美景，我们除了精心保护便什么都不能做了，这样，这一古老的建筑就可以保留到百年之后。

21世纪的"桩住宅"
大约5年前，一群朋友和同事来到了Gamboa，打算将一个"桩住宅"打造成四部分，一种可以朝向街道和大海的阁楼式住宅。之后，另外一个人买了其中一部分，自己将其改造成了酒店。随后，一对年轻夫妇买了另外一部分，并改造成小旅馆，主要接待欧洲游客。至此，这一住宅区便呈现出一个新的姿态。正是在这个背景下我们受邀设计一个带有12间卧室的精品酒店。

建筑改造提案
原有的"桩住宅"均为一层式结构，因此显得格外地宽敞。其位于污水排水系统之下，桩柱已严重变形，无法修复，因此不容易保留下来。值得欣慰的是，原有的山达木木板及当地木材打造的门窗可以被保留下来，从建筑或历史元素来看，这一独特的建筑不属于任何一种风格和时代。

之后，必须获得相关部门的许可。首先，须向当地海事部门申请；然后，查看当地法律规定的建筑条件——建筑必须采用木材搭建，必须使用桩柱支撑，带有规格为2米X5米的前院，桩柱上的结构必须向后凹陷便于加建一层。

以"缺口"为出发点
我们确定建筑延伸到海中的最大长度不能超过本街区最长的建筑本身，同时我们决定将

In the year 1960, a major earthquake and seaquake destroyed most of the Palafitte neighbourhoods that existed in other cities, like Ancud, Quemchi, Dalcahue and Quellón. The force of the wave swept them away, making them disappear from the edge of cities and towns. Only those in Mechuque and Castro were saved. In this last city because by being in the centre of the island and protected by their fjord, the wave arrived attenuated and only flooded them. The job afterwards was to lift them up, and so happened with the railways making the train disappear.

Palafitte Defense
The idea of eradicating them seemed to us as an aberration, so we embarked on their defense and in the document Letter for Chiloé we proposed that the real issue was the necessity to improve them, to provide them with basic services since they were part of the cultural identity of Chiloé, and their inhabitants, heirs of an architectural tradition.

In the middle of the debate, one night the town wakes up with the sound of fire sirens down Blanco Street. The Palafitte barracks are on fire. Little can be done to save these combustible constructions and their contents.

At the end, Palafittes on Pedro Aguirre Cerda and Lillo streets are eradicated, and their inhabitants relocated in a different neighbourhood where some of them still make boats in their small backyards, longing a life by the sea.

On our side we continued to develop public works at Taller Puertazul that reiterated the use of the Palafitte condition, such as Dalcahue's fair, which entered the sea and conciliated territory and maritory in one single concept.

Urban - Architectural Renovation
It was at about that time that we got the commission to design a restaurant, Don Octavio, that had obtained a maritime concession across from Unicornio Azul Hotel, where once were located the Palafitte barracks. We proposed it as a wooden seashell.

However, the new authorities banned permission for new Palafitte constructions to privilege the existence of a shore drive that would allow seeing the exuberant beauty of this austral landscape.

At the beginning of the 1990s, a group of Spanish architects, from Architects Without Borders, aware of the architectural and cultural value of these buildings, presented a project to the European Economic Community to get funds for their restoration, and managed to get the Municipality to act as counterpart.

There were two offices doing the works. We worked at Gamboa neighbourhood and proposed that considering the resources available, what should be done as intervention was to incorporate bathrooms in each house, and rebuild the façade facing the sea.

While performing this project, Chile's forestry institute contacted us with the Swedish company FALUN with which we developed a painting project for the Palafittes of Gamboa, using their colour palette, with over three hundred years of tradition.

其与周围的结构分离开来，中间的"缺口"空间（原被用于储存木柴或通向大海的空间）与景观紧密相连。

室内，我们以一条纵向"缺口"——走廊及两条横向通道，布置空间格局。走廊将入口大厅与大海相连，两条横向通道与周围建筑相通。大堂朝向街道，走廊引向起居室、餐厅、厨房。走廊的一侧是服务台，另一侧则是通往上层、洗衣房及办公区的楼梯。一层设有三间卧室。二层采用同样的格局布置，其余的卧室便全部位于这里。三层包括一个多功能空间，带有大露台。尽管设计与传统的样式脱离，但却使其高于周围的建筑与景观，光线从玻璃地面经由天窗引入到二层走廊内。

可持续发展
从节能角度来讲，这一建筑结构紧凑，中央仅需要走廊连通，并与周围的结构相协调。

从技术层面来讲，这一建筑秉承了百年传统，充分运用当地木材、旧建筑回收结构。更为重要的是，当地最具特色的木工工艺被重新使用，使其成为智鲁岛建筑的顺延。

另一方面，墙壁与地面分离开来的做法，以便达到最佳热能消耗及隔音效果。双层玻璃窗的安装、屋顶的沥青膜以及起居室内的木材锅炉（供应整幢建筑的热水）均起到了一定的节能作用。

室外，我们主要选用当地材料；室内则运用由当地木材打造的工业板；灯光以及装饰物的运用则更加突出了小岛文化特有的毛绒、织物及木材等。

标准的"缺口"与新时代的走向
毋庸置疑，我们的这一设计弥补了标准的"缺口"，"桩住宅"不仅仅用于拍摄，更可用来居住。在那里可以感受潮汐的变化，欣赏飞过的天鹅，清晨在翠鸟敲打窗户的声音中，或是烤箱中面包的香味中醒来，品味当地的美味食品。

爱德华·罗哈斯
2011年6月写于卡斯特罗

The refurbished Palafittes, connected to basic services in the city, were now prepared to assume a new image facing the 21st century.

Patrimonial Value

At this point it was clear to us that Palafittes are the most accurate expression of a wood architecture that in Chiloé, as well as its culture, are in permanent change. This was very clear seeing the antique photographs of these neighbourhoods, and the ones we took by then and have somehow become antique. The grain, the texture, the story, the wood, the embroidered border, was always the same, a part of their immanent value. Among them, there was the interstice left between the different volumes used to enter firewood, or to see through and glance at the landscape that filters through them enhancing the condition of the place in the urban world.

Making Heritage as Own

At this point of the story, it is important to comment that eighteen years ago I was offered to buy the doors and windows of an old Palafitte that stood alone on the shore and was about to be demolished for having completed its useful life. With my family, we decided to buy it and inhabit it, and with this we proved the fact that restoration of an antique wooden building involves a strong commitment and resources.

But greater was our fascination for a place where sea gets to the level of the terrace during high tides and over fifty different species of birds come to feed on small seashells and crustaceans when the tide falls.

And so, faced with this paradise, we could not do anything different but to care about it so that new generations can come and know what architecture one hundred years ago was like.

21st Century Palafitte

About five years ago, a family of friends and colleagues embarked on the project of converting a Palafitte in the neighbourhood of Gamboa, into four departments, a sort of lofts with both view to the street and to the sea, proving that there was great interest in living in these locations. Then, another person bought a small one, and without the help of architects as he told me in person, rebuilt it as an apart-hotel. Later, a young couple bought another one and transformed it into a Hostel, destined mainly to European tourists, and with the construction of a twin Palafitte to the one with the four departments, it was clear that this neighbourhood is assuming a new destiny and it is in this context that we received the request to develop the project of a twelve-bedroom boutique hotel at 1326 Ernesto Riquelme Street.

An Architectural Renovation Proposal

The existing Palafitte was a quite wide, one storey house, and it was below the level of the sewer system, with stilts in bad shape, which means it was not possible to repair it, even less to restore it. At the most it was possible to remove its millenary alerce shingles that were in good conditions, and recycle some doors and windows made of native wood. This particular building was not representative of any determined style or time, in architectural and patrimonial terms.

And then, some intervention parameters had to be defined. The first thing is that along with the project, it was necessary to start the immediate request for a maritime concession of the site, which was something that no one had in the neighbourhood. Not even the previous interventions. Then there were the building conditions established by law. The building had to be made of wood, standing on stilts, with a front yard of at least two metres and a maximum height of five metres, above which the volume should be recessed in order to build an additional floor.

Start Based on Interstices

We defined that the maximum length of the building's penetration into the sea could not be superior to that of the longest Palafitte existing, and we established an emplacement proposal, separating it from the neighbouring Palafittes generating the interstices that put us in contact with the landscape, traditionally used to enter firewood or go down to the sea.

Inside, we assembled the building through the use of one longitudinal interstice - hallway and two transversal interstices to organise the project. The first one communicates the entrance foyer with the sea, and the other two with the Palafitte neighbourhood itself, with its walls and shingle textures, wood and metal plates.

We put the lobby toward the street and the interstice - hallway conducts us to the living room, dining room, collective kitchen, which facilitates the encounter among guests. One side of the hallway hosts the service and on the other side there is the staircase leading to the second floor, the laundry and the offices. This first floor counts with three comfortable bedrooms.

On the second floor the same structure is maintained, with the rest of the bedrooms, and on the third floor we created a multi-purpose space with a large terrace, which even though it does generate a rupture with the traditional typologies, it makes it possible to be above the neighbourhood and the landscape, making a true statement of living the fifth façade, which also filters into the building through the glass floor that converts into skylights at the end of the hallways on the second floor.

The Sustainable Condition

In technological terms, the building continues a centennial tradition through the work with native wood, recycling pieces from the old building, and above all, the carpentry tradition that converts it into a work sustained by the genetic codes of the architecture of the place, and therefore, it becomes part of the architectural continuum in Chiloé.

In terms of energy efficiency, the building is presented as a compact volume, broken only by the interstice hallways, which dematerialise the volume integrating its grain to the rest of the Palafittes in the neighbourhood.

On other side, its walls and floors are isolated to achieve the best thermal and acoustic efficiency, incorporating double glazed windows installed in hermetic PVC windows, as well as asphalt membranes on the roof to allow the construction of terraces, and a heating system from a wood boiler located in the living room, from which hot water radiators are connected around the entire building.

Though outside we worked with local traditional materials, inside we used industrial plywood made of native wood, using lighting and interior design to enhance the value of the characteristic materials of the insular culture such as wool, fibers, and of course, wood.

Paradigm Rupture and Openness to New Times

Without a doubt with this work inserted in the neighbourhood and its scale, we generated the rupture of a paradigm. The one that Palafittes are not only for the photographs but to be inhabited, to feel the flow of the tides, experience the approach of black-necked swans displaying the splendor of their plumage, and to be awakened in the morning by a kingfisher, knocking with its peak on the window, or by the smell of homemade bread fresh from the oven, to enjoy with jam, coffee and dairy products from the area, and feel the experience of living in a Palafitte, to be there, as we once said, sitting at the stalls of cosmos, of a cultural cosmos looking for its place and time in the world, from the valuation of its singularities and identity.

Edward Rojas
Castro, June 2011

Hollywood Roosevelt Hotel

罗斯福酒店

Location: Los Angeles, California, USA
Completion: 2011
Designer: Studio Collective
Photographer: Thompson Hotels
地点：美国 加利福尼亚州 洛杉矶
翻新时间：2011年
设计：集合工作室
摄影：汤普森酒店集团

BACKGROUND

Since 1927, the Hollywood Roosevelt Hotel has been the playground of luminaries including Clark Gable, Carole Lombard and Marilyn Monroe and the birthplace of the Academy Awards. As Hollywood continues to redefine itself, so does its landmark hotel, offering the ultimate luxury guest experience in-room and out, while paying homage to its roots. With the addition of the new spaces to the hotel's existing portfolio of celebrated venues, the Hollywood Roosevelt Hotel reaffirms its status as a social epicentre for Hollywood's elite.

OBJECTIVE

Like any classic Hollywood star, Thompson Hotels' Hollywood Roosevelt Hotel has continuously reinvented itself over the years, and the iconic 300-room hotel has once again seen a rebirth with the redesign of the historic Cabana Rooms and the debut of three innovative entertainment venues: Public Kitchen & Bar, The Spare Room and Beacher's Madhouse.

SOLUTIONS

Public Kitchen & Bar:
Tim Goodell and The Domaine Restaurant Group introduce an exciting new dining experience at the Hollywood Roosevelt Hotel with the opening of Public Kitchen & Bar. A complete transformation in food, design and experience is about to take hold in the iconic Hollywood Roosevelt Hotel lobby, in the space formerly known as the Dakota Steak House. Per Goodell, Public Kitchen & Bar will be about community: a bustling, comfortable vibe that inspires social interaction, along with great food, cocktails and vibrant music. The overall design will celebrate the landmark hotel's original architecture & Spanish Colonial architecture. The restaurant will be highlighted by a new axis

of entry via the restaurant's centre - a large, masculine, three-sided, marble-topped bar will look out to the hotel's grand lobby and serve as the entrance point. The airy and boisterous dining room will feature brass chandeliers anchored by the hotel's original 1927 ceiling fresco, recently uncovered and restored. Rustic oak cabinetry, black walnut tables and cabernet leather booths round-out the lively and comfortable vibe.

The Spare Room:
The Spare Room is the newest addition to the Hollywood Roosevelt property, located above the lobby, and overlooking Hollywood Boulevard. Providing unparallelled amenities to hotel guests as well as a unique destination for Hollywood's architects of influence, this modern-day parlour highlights the camaraderie and spirit of gaming with two reclaimed bowling lanes, and custom-built backgammon tables. Combining cocktail and culinary culture along with exceptional service, the Spare Room is a social experience where details (and taste) matter.

Beacher's Madhouse:
The Hollywood Roosevelt Hotel welcomes Beacher's Madhouse is a revolutionary Vaudeville-inspired theatre development on the lower level of the infamous hotel, with European influences and echoes of the Folies Bergére. Guests will enter the theatre through an inconspicuous passageway, hidden behind a library bookcase, and travel through a light tunnel from ordinary existence into a fantasy-filled world.

The venue extends 2,500 square feet, featuring tiered glass stages, oversized objects, and floor-to-ceiling crystal chandeliers. The walls are embedded with a mish-mash of unique objects, such as mannequins and old pieces of furniture, bringing the setting to life on an unprecedented level.

1. Lobby before renovation
2. Spare room
1. 翻新前大堂
2. 休闲区

3

背景

自1927年以来，罗斯福酒店一直是举办各种知名活动的首选场所，并见证了奥斯卡的诞生。好莱坞的不断发展促使着这一地区标志性酒店的改变，旨在为客人提供奢华而又不失温馨的居住体验。新建的空间更赋予酒店全新的功能——好莱坞精英人士的集散地。

目标

汤普森旗下的罗斯福酒店可堪称"好莱坞之星"，近年来不断进行完善。300间客房重新设计，同时更增添了三个全新的娱乐空间：公共厨房与酒吧、客厅及剧院。

过程

公共厨房及酒吧：

这一空间取代了原来的牛排屋，无论在食物、装饰及体验上都为酒店客人带来全新的感觉。整体设计突出了酒店原有的西班牙殖民风格，一条新的中轴线穿越餐厅的中央区。宽大的大理石饰面吧台摆放在入口，从这里可望向酒店宽敞的大堂。风格轻快的就餐区内洋溢着热烈的氛围，黄铜材质吊灯从画有壁画（1927年设计）的天花上垂下来，橡木家具、黑色胡桃木餐桌以及皮质座椅更是别具特色。

客厅：

新建的客厅位于大堂正上方，可俯瞰好莱坞大道的美丽景致。这里有保龄球道、定制的西洋双陆棋棋桌，突出营造休闲娱乐的氛围。鸡尾酒、丰富的饮食以及更多的特色服务给这里增添了别样的体验，而注重细节则构成了设计的亮点。

剧院：

这一空间布置在酒店地下室内，其设计灵感源于歌舞杂耍表演，并受到欧洲风格的影响，在这里可以找到法国著名剧院"Folies Bergére"的影子。经过一条隐藏在图书室书架后面的走廊以及光亮的隧道便可进入到这里梦幻的世界。

剧院面积达232平方米，阶梯状的玻璃舞台、超大号的饰品以及水晶吊灯构成了主要特色。墙壁上嵌满各种独特的小物品，如人体模型、老家具等，赋予空间独特的生命力。

3. Public kitchen and bar
4. Lobby and restaurant
5. Private dining
3. 开放式厨房及酒吧
4. 大堂及餐厅
5. 私人餐厅

6. Teddy's and library bar
7. Beacher's madhouse
8. Lounge in the spare room
6. 酒吧及小图书室
7. 俱乐部
8. 休闲区内休息室

9

9. Upstairs bedroom
10. Tower room
9. 上层客房
10. 顶层客房

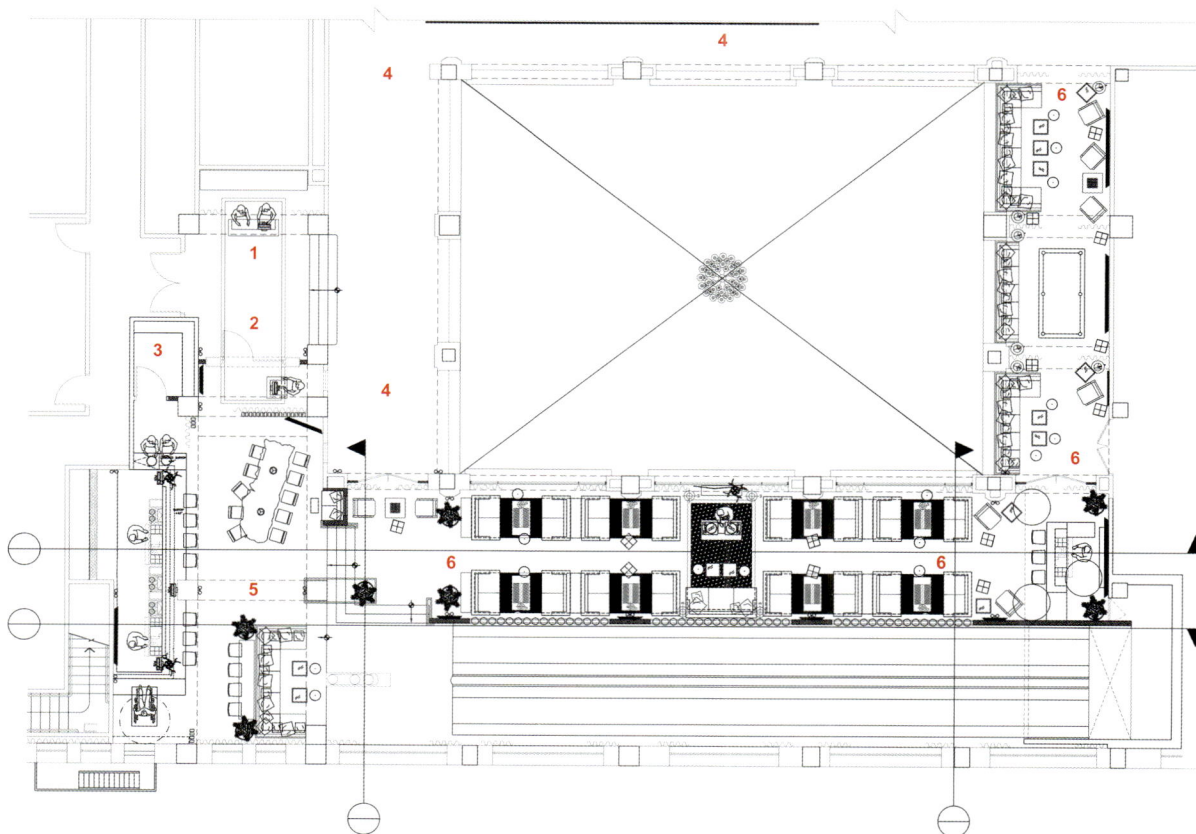

1. Reception
2. Foyer
3. Storage
4. Mezzanine
5. Bar area
6. Salon

1. 接待台
2. 大厅
3. 储物区
4. 中层区
5. 酒吧
6. 沙龙

Sofitel "The Grand", Amsterdam

阿姆斯特丹索菲特大酒店

Location: Amsterdam, the Netherlands
Completion: December 2010
Architect: Boparai Associates BV architekten bna
Interior Design: SM Design
Photographer: Glenn Aitken
Gross Floor Area: 22,660m²

地点：荷兰 阿姆斯特丹
翻新时间：2010年12月
建筑设计：Boparai 建筑设计公司
室内设计：SM 设计公司
摄影：格伦·艾特肯
建筑面积：22,660平方米

BACKGROUND

Hotel "Sofitel Legend Amsterdam, the Grand", is located between two canals in one of the oldest parts of the city of Amsterdam. The hotel consists of a complex of six different buildings which together form an ensemble. A major part of the complex is listed as a historical monument. Since the 15th century the site has housed various functions: a convent, an inn, the headquarters of the Dutch navy, the city hall and now a hotel. Some of the existing buildings date from the 17th century when they housed the admiralty. In 1992 the whole complex was converted from the Amsterdam city hall to a luxury hotel. The architectural styles today include the renaissance style, neo-classicism, and the typical Amsterdam School style of the early 20th century.

OBJECTIVE

Some years later the present owners decided to do a major refurbishment in order to justify their ambition to make this hotel the flagship of the hotel group.

SOLUTIONS

The refurbishment was completed in 2010. The challenge was to alter the hotel in such a way that the essence of the historical character was respected, while at the same time the existing logistics, routing, fire safety and appearance of the interior spaces was improved and renewed.

For example the meeting spaces which could previously only be accessed via the open courtyard are now internally accessible, while other problems such as the inadequate access to some of the best rooms, the layout of the kitchen and restaurants, etc. have been addressed and solved, so that the hotel functions efficiently for both guests and staff. The interiors have been carefully designed to combine modern facilities with the historical architectural styles in a harmonious way.

Award name:

Venuez Hospitality & Style Awards 2010, Hotel Chain Design of the Year and Restaurant Design of the Year

Villegiature Awards, Best Hotel Interior in Europe

World Travel Awards, the Netherland's Leading Hotel 2010

Expedia Insiders' Select 2010, chosen as one of the world's top 1% of hotels and best hotel in Amsterdam

1. Reception before renovation
2. Restaurant before renovation
3. Imperial suite before renovation
4. Façade at dusk
1. 翻新前接待区
2. 翻新前餐厅
3. 翻新前帝王套房
4. 黄昏下的酒店外观

1. Entrance
2. Lobby
3. Reception
4. Office
5. Meeting
6. Flower shop
7. Staff cantine
8. Security
9. Parking
10. Swimming pool
11. Business
12. Back of house
13. Courtyard prinsenhof
14. Garden courtyard
15. Restaurant
16. Kitchen area
17. Goods

1. 入口
2. 大堂
3. 接待台
4. 办公区
5. 会议区
6. 花店
7. 员工就餐区
8. 安全通道
9. 停车场
10. 游泳池
11. 商务中心
12. 后台服务区
13. 庭院
14. 花园
15. 餐厅
16. 厨房
17. 货物储存区

5

背景

阿姆斯特丹索菲特大酒店坐落在古城内两条静静的运河之间，由六幢建筑组成，其中一部分建筑已被列为历史文物。自15世纪建成之初，这一建筑便行使着多样的功能——会议中心、旅馆、荷兰海军总部、市政厅及酒店。1992年，整栋建筑被改建成奢华酒店，其建筑风格更是多样化，文艺复兴风格、新古典主义风格、阿姆斯特丹特色一应俱全。

目标

如今，酒店现有主人决定对其进行修缮以使其成为酒店界的旗舰店。

过程

修缮工作于2010年完成。在这之中，设计师面临的主要挑战即为在保留建筑原有历史特色的情况下将室内通道、消防设备及空间等进行翻修。

具体的工作可以举个例子来说明。起初，会议室只能通过室外庭院进入，改建之后可从建筑内部进入。其他如通往不同房间的入口也增添了一些，厨房及餐厅的格局也被改变，为酒店客人及员工带来方便。室内空间经过仔细设计，现代化的设备与传统将建筑特色完美融合。

获奖

2010年度酒店及餐厅设计"Venuez Hospitality & Style"大奖；
欧洲最佳酒店设计"Villegiature"奖；
2010年度世界旅行奖"荷兰杰出酒店"奖；
2010年度"世界1%顶级酒店"奖。

5. Lobby
6. Reception
7. Restaurant
5. 大堂
6. 接待区
7. 餐厅

8. Living room of the imperial suite - the main theme of simplicity dominating every corner
9. Fireplace performing as the focal point in the library and creating a warm atmosphere
8. 帝王套房内客厅——简约是整个空间设计的主题
9. 壁炉成为图书室的中心装饰，营造了温暖的氛围

10. Orange curtain, purple ceiling and green water making an elegant world in the spa
11. Character sketch on the wall of the suite adding touches of fun to the whole space
12. The Imperial Suite Maria de Medici featuring wall papers embedded in the wooden frame and decorative object on the wall
10. 橘色的窗帘、淡紫的天花以及浅绿的池水构成了一个典雅的水疗空间
11. 套房墙壁上的人物素描画增添了空间趣味性
12. 帝王套房内镶嵌在木框内的壁纸以及墙壁上的饰品成为焦点

Hotel Novotel Amsterdam

阿姆斯特丹诺富特酒店

Location: Amsterdam, the Netherlands
Completion: May 2010
Architect: Boparai Associates BV architekten bna
Interior Design: Proof Consultancy Ltd.
Photographer: Luuk Kramer
Area: 14,900m²
Construction Area: 30,500m²
地点：荷兰 阿姆斯特丹
翻新时间：2010年5月
建筑设计：Boparai 建筑设计公司
室内设计：Proof 咨询有限公司
摄影：卢克·克雷默
占地面积：14900平方米
建筑面积：30500平方米

BACKGROUND

The Novotel was built in 1971 as the Alpha Hotel nearby the RAI Exhibition and Convention Centre, the heart of Amsterdam's financial district. With 611 rooms it is the biggest hotel in Amsterdam. The appearance of the hotel was outdated and the routing, logistics and public spaces did not meet the current requirements of the hotel.

SOLUTIONS

On the ground floor level, approximately 1,500 m² of congress facilities have been added and the whole interior renewed. The routing between reception lobby and the food and beverages areas have been optimised and made more accessible, by removing existing barriers and by re-routing elements, such as escape routes. The terrain layout has been changed, so that guests are guided to the main entrance instead of the rear, as was previously the case. The previous rear entrance is now a specialised entrance for groups.

Although the original façades of the bedroom block still meet thermal requirements, they looked very worn out and dated. To rejuvenate these façades, without demolition of the existing concrete façade and also without adding too much weight to the structure, a relatively light glass façade has been chosen. The glass panels are silk screened with a light pattern, so that the façade changes texture with the changes of light during the day.

Both the existing and the new façades on the ground floor level are cladded with light weight fibre reinforced concrete panels, so the entire ground floor gets a uniform architectural expression.

The interior of the building has been closely coordinated with the exterior, in order to express the light, warm and welcoming character of this hotel.

1. Aerial view before renovation
2. Façade before renovation
3. Contrast between before and after renovation of façade
4. General view
5. Entrance and sign
1. 酒店翻新前鸟瞰图
2. 酒店翻新前外观
3. 酒店外观翻新前后部分对比
4. 酒店建筑全景
5. 酒店入口及标识

6. Reception and lobby
7. Lobby
8. Convention centre
6. 接待台及大堂
7. 大堂
8. 宴会厅

背景

诺富特酒店始建于1971年，之前被称作"Alpha酒店"，位于阿姆斯特丹金融区中心，与RAI国际展览中心相邻。作为阿姆斯特丹最大的酒店，其共有611间客房。酒店外观已完全过时，室内通道及公共空间等已完全不能满足现有需求。

过程

整个室内空间进行重新装饰，一层增添了大约1500平方米的会议区。接待大厅同餐饮区之间的障碍被彻底摒弃，逃生通道等元素加以优化，使其更加人性化。通往酒店的线路被改变，之前客人需通过后门进入，如今可直接到达主入口，后门入口被用作团体通道。客房区原有外观虽能满足热能保温需求，但已破旧不堪，完全过时。为此，设计是将外观进行翻新，原有的水泥层被保留下来，在此基础上加建了一层玻璃结构，以减少建筑整体负重。玻璃板经过丝印技术处理，随着光线的变化，表面的图案会不断变化。一层原有外观及玻璃外观处均采用纤维土板镶嵌，从而形成统一的视觉效果。

酒店的内部设计与外观相互呼应，突出其明亮、温馨及友好的氛围。

9. Black counter and white decorative objects suspended from the ceiling contrasting with each other in the bar
10. Coffee Corner with green as the main decorative colour to create fresh and lively feeling
11. The wooden panels furnishing the buffet counter as well as the unfinished brick wall highlighting the natural feature
12. Restaurant
13. Different colours converging in the restaurant forming a gorgeous space

9. 酒吧内黑色的吧台与屋顶上垂悬下来的白色装饰物在色彩上形成鲜明对比
10. 绿色用来粉饰咖啡区，营造清新活跃的气氛
11. 餐厅内自助餐台上的木板装饰以及未经修饰的砖石墙壁突出了自然特色
12. 餐厅
13. 餐厅内各种艳丽的色彩融合，打造了一个缤纷的空间

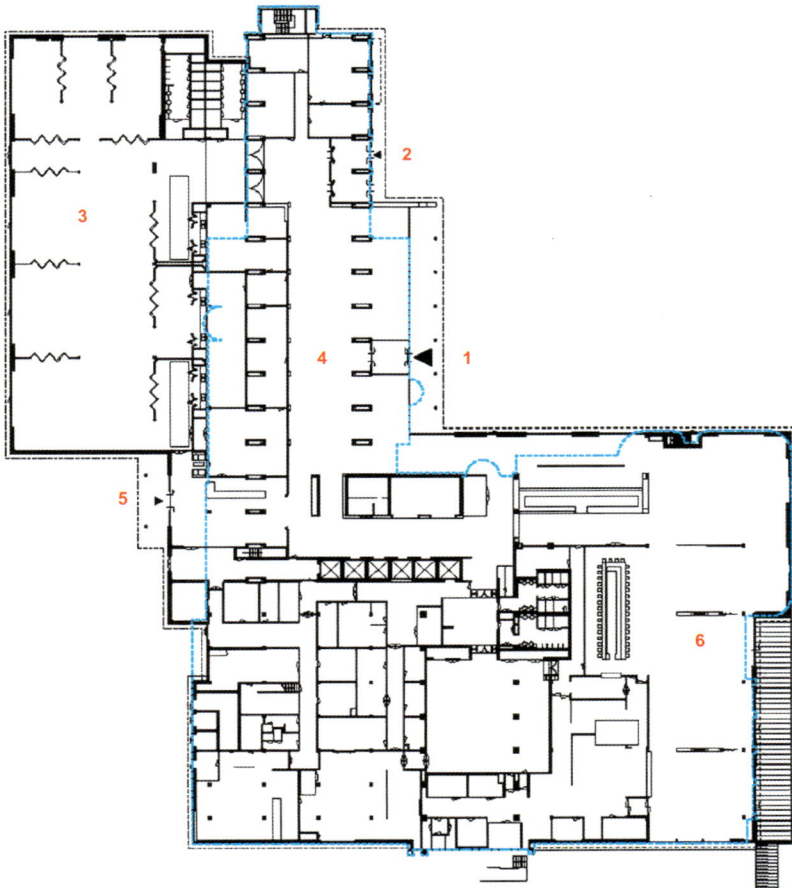

1. Main entrance
2. Convention centre entrance
3. Convention centre (new)
4. Lobby area
5. Group entrance
6. Restaurant

1. 主入口
2. 宴会厅入口
3. 新宴会厅
4. 大堂
5. 团队入口
6. 餐厅

Mandarin Oriental Jakarta

雅加达文华东方酒店

Location: Jakarta, Indonesia
Completion: 2010
Designer: Lim.Teo+Wilkes Design Works Pte Ltd.
Photographer: Marc Gerristen

地点：印度尼西亚 雅加达
翻新时间：2010年
设计：LTW 装饰设计有限公司
摄影：汤马克

BACKGROUND

The hotel is situated in the heart of cultural and historical part of Jakarta. And it is also close to the central business district. This makes the hotel a wonderful location for both tourists and business travellers to stay. The hotel needed a new fresh look badly as it has not been renovated for many years and especially with competition like newly opened Grand Kempinski Jakarta and newly renovated Grand Hyatt Jakarta - all within walking distance.

OBJECTIVE

LTW's design is to transform this hotel to meet the demands of a modern traveller be it a tourist seeking to discover the charms of this Dutch colonial influenced city or a businessman seeking his fortune in this bustling city centre. LTW Design Works then set to give the hotel a contemporary updated look while retaining the local cultural essence. The idea is to give the hotel guests a calm and serene space during their stay in bustling Jakarta.

SOLUTIONS

Transformation of the hotel interiors took about one year. Design development process was about one year. This project presented two major constraints - budget and existing building structure. The building is old with ceiling heights, old mechanical and electrical systems, etc.

Jakarta is well known for its traffic jam - choked roads and highways in the tropic heat and humidity. So the new design is to welcome the guests with comfortable room and equally spacious bathroom - all with modern furnishings, fittings and trappings of current technology. And yet at the same time to let the guests experience a sense of place - of Indonesia - by bringing in the cultural elements of Indonesia and the locality into the hotel.

LTW worked very closely with the operator to fully understand its needs. The most dramatic makeover is the lobby - from a dark interior to a bright iconic space. One can see this upon arrival at the hotel - metal cut screen behind reception counter at lobby and the intricately hand-carved wood screen at the ballroom. These are traditional motifs found in Indonesian arts and crafts. Instead of having a typical hotel lobby chandelier, LTW playfully adorned the structural column in the lobby with precisely-cut diamond-shaped mirrors. The column is miraculously transformed into an iconic "chandelier" - a dazzling mirror column playfully reflecting the hotel interiors.

All guestrooms are updated to current standards. Most furniture pieces were made locally with imported fabrics and leathers for upholstery. Specially selected pieces of artwork, reflecting local Indonesian flavours, are installed in the guestrooms and suites. Rich textiles and leather are used throughout the hotel to add the warmth and luxe.

1. Lobby before renovation
2. Restaurant before renovation
3. Guestroom before renovation
4. Renewed lobby
1. 酒店翻新前大堂
2. 酒店翻新前餐厅
3. 酒店翻新前客房
4. 翻新后大堂

5

背景

该酒店坐落于雅加达文化及历史中心区域，并与商业区相邻，因此，这里是旅行者及商务人士的首选住宿之所。酒店多年来未进行过翻修，与周围新开的凯宾斯基酒店及新装修的君悦酒店相比，略显逊色，因此急需进行翻新。

目标

设计师致力于使酒店既能满足旅行者的需求，让他们尽情去发现荷兰殖民地的魅力，同时又能适合商务人士的口味，让他们在繁华的城市中心探寻属于自己的"宝藏"。设计旨在赋予酒店全新的现代风格外观，同时保留当地特有的文化精髓，为客人营造一个恬淡而不失豪华的住宿氛围。

过程

酒店室内改造工作花费了将近一年的时间，设计规划也花费了一年时间，遇到的主要问题即为预算短缺及建筑原有的低矮屋顶、破旧的机械及电力系统等。

众所周知，雅加达的交通拥堵状况异常严重，拥挤的马路加之热带地区的湿热天气，往往让人感觉压抑。因此，东方文华酒店的翻新将主要为客人营造舒适的环境——温馨的客房、宽敞的浴室、现代风格的家具一应俱全。同时，设计还注重让客人体验印度尼西亚的地域特色，将该区的文化元素与特色风格融入到酒店之中。

设计师与酒店管理者密切合作，便于全面了解客人的需求。酒店中最大的变化即为大堂的改造，从灰暗低矮的空间转变成明亮宽敞的场所。步入酒店，大堂接待台后面的金属切割屏风以及舞厅内的手工雕刻木质屏风别具特色，这些传统的花式是印度尼西亚艺术及手工艺品的精髓。此外，设计中摒弃了传统的枝型吊灯，而是在大堂柱子上安装水晶形状的镜子，构成了独特的灯饰。

所有的客房更是达到现代风格标准，家具采用进口织物及皮革等打造，增添了温馨与奢华。精心选择的艺术作品彰显当地特色。

5. Chinese restaurant
6. Presidential suite living
7. Presidential suite dining

5. 中餐厅
6. 总统套房起居室
7. 总统套房餐厅

8

Presidential Suite Layout
1. Entry
2. Bedroom
3. Bathroom
4. Walk-in closet
5. Powder room
6. Pantry
7. Dining
8. Living
9. Study
总统套房平面
1. 入口
2. 卧室
3. 浴室
4. 步入式衣柜
5. 化妆间
6. 配餐室
7. 餐厅
8. 起居室
9. 书房

8. French windows in the bar bringing outdoor landscape inside
9. The carpet corresponding with the sofas and chairs in colour and the flower patterns on it adding vividness to the whole bar

8. 酒吧内，落地玻璃窗将室外美景引入进来
9. 酒吧内，地毯在色彩上与沙发及座椅相互呼应，上面的花朵图案则增添了趣味性

10

11

10. Guestroom
11. Spacious bathroom
12. Presidential suite bedroom
10. 客房
11. 宽敞的浴室
12. 总统套房卧室

Fairmont Peace Hotel

和平饭店

Location: Shanghai, China
Completion: July 2010
Designer: Ian Carr and Connie Puar / HBA
Photographer: Courtesy of Peace Hotel
Area: 36,317m²
地点：中国 上海
翻新时间：2010年7月
设计：伊恩·卡尔，康妮·普尔（HBA 设计顾问公司）
摄影：和平饭店
面积：36317平方米

BACKGROUND

Once, the Fairmont Peace Hotel was The Cathay Hotel and it was the most glamorous hotel of 1930s' Shanghai. It stood beside the Huangpu River, at the corner of the Bund and Nanjing Road, the most famous address in Shanghai.

The Cathay Hotel was the enduring vision of Sir Victor Sassoon, Shanghai's exotic property and finance tycoon, who had a passion for racehorses, high style and high society, and for giving fabulous parties and extravagantly flamboyant costume balls.

SOLUTIONS

By the 21st century Shanghai was host to 2010 World Expo and the famous hotel on the Bund re-opened as the Fairmont Peace Hotel.

There have been three years of careful, extensive restoration and the new Fairmont Peace Hotel brings to Shanghai the very best of contemporary sophistication. While much of the old world has gone, much remains. There are still the 1920s Lalique glass ornaments and fixtures, especially designed by the Lalique workshops in France. The original Art Deco frieze of the two greyhound designs, the emblem of the hotel, still surrounds the entrance atrium, and once again light pours through the original skylight, gilding all who stand below in the golden light of energy, optimism, and success.

The imposing granite statue of Marshall Chen Yi, PLA liberator of Shanghai, first Foreign Minister of PRC, first Mayor of Shanghai, gazing across the city, stands on the other side of the Bund opposite the hotel entrance. The grand revolving doors once more open onto the Bund, still the most famous address in Shanghai, and the windows overlook the Huangpu River, "mother River of Shanghai", forever flowing out to all the world.

It shimmers with the years that have made history. The green copper pyramid of the Fairmont Peace Hotel stands as witness to the style and energy of Shanghai and to the promise of China.

1. Entrance hall facing the Bund in the 1930s
2. Original Indian Suite
3. Original Jacobean Suite
4. Main entrance
1. 20世纪30年代朝向外滩的入口大厅
2. 翻新前 "印度" 套房
3. 翻新前 "詹姆士一世" 套房
4. 饭店主入口

背景

和平饭店原名为华懋饭店，在20世纪30年代的上海是最负有盛名的酒店。它坐落于黄浦江之滨，位于外滩与南京路的交界口，它的所在是上海最为著名的地址。

华懋饭店是其缔造者维克多•沙逊爵士的愿景。沙逊爵士在当年的上海滩是一位令人瞩目的资产家和金融大亨，热衷于赌马和奢侈高端的社交生活，他一手将华懋饭店打造成当时上流社交派对和奢华晚宴舞会的流光溢彩之地。

过程

跨入21世纪，上海成功举办2010年世博会，而这座外滩历史名店也成为费尔蒙旗下的地标性酒店，展开盛世新纪元。

历经3年的大规模精心修复，和平饭店重新焕发荣耀光彩，成为上海最负盛名的一道风景。昔日的浮华虽已淡去，却未带走其留下的痕迹。和平饭店仍然保留了20年代的拉力克（Lalique）水晶玻璃浮雕嵌饰，这是当时特别向法国拉力克水晶玻璃工坊定制的。酒店原有的标志、由装饰派艺术一双灰狗图案组成的饰带仍然环绕于酒店入口处的中庭墙顶。华丽的光晕，再一次穿越天窗倾泻而下，将立于厅堂之上的人们笼罩在带来活力、乐观和成功的金色光芒中。

庄严的陈毅塑像伫立于花岗岩底座之上，屹立于外滩江滨，他是上海解放之战的总司令，新中国的第一任外交部长和第一任上海市长，他凝视着上海，守望着和平之门。和平饭店的旋转大门又一次在外滩开放，依旧是沪上声名最盛的地标；和平之窗俯瞰"上海的母亲河"黄浦江，眷眷之流奔向五湖四海。

历经数十载风云沧桑，依然闪耀着历史的光芒。和平饭店的铜质尖顶，见证着魅力四射的上海，见证着前景远大的中国。

1. Jazz bar
2. Lobby lounge
3. Public toilets
4. Circulation
5. Boou sttop
6. Deli
7. Hotel lobby
   / reception
8. Atrium

1. "爵士"酒吧
2. 大堂休闲廊
3. 公共卫生间
4. 通道
5. 博物馆
6. 蛋糕房
7. 大堂／接待台
8. 中庭

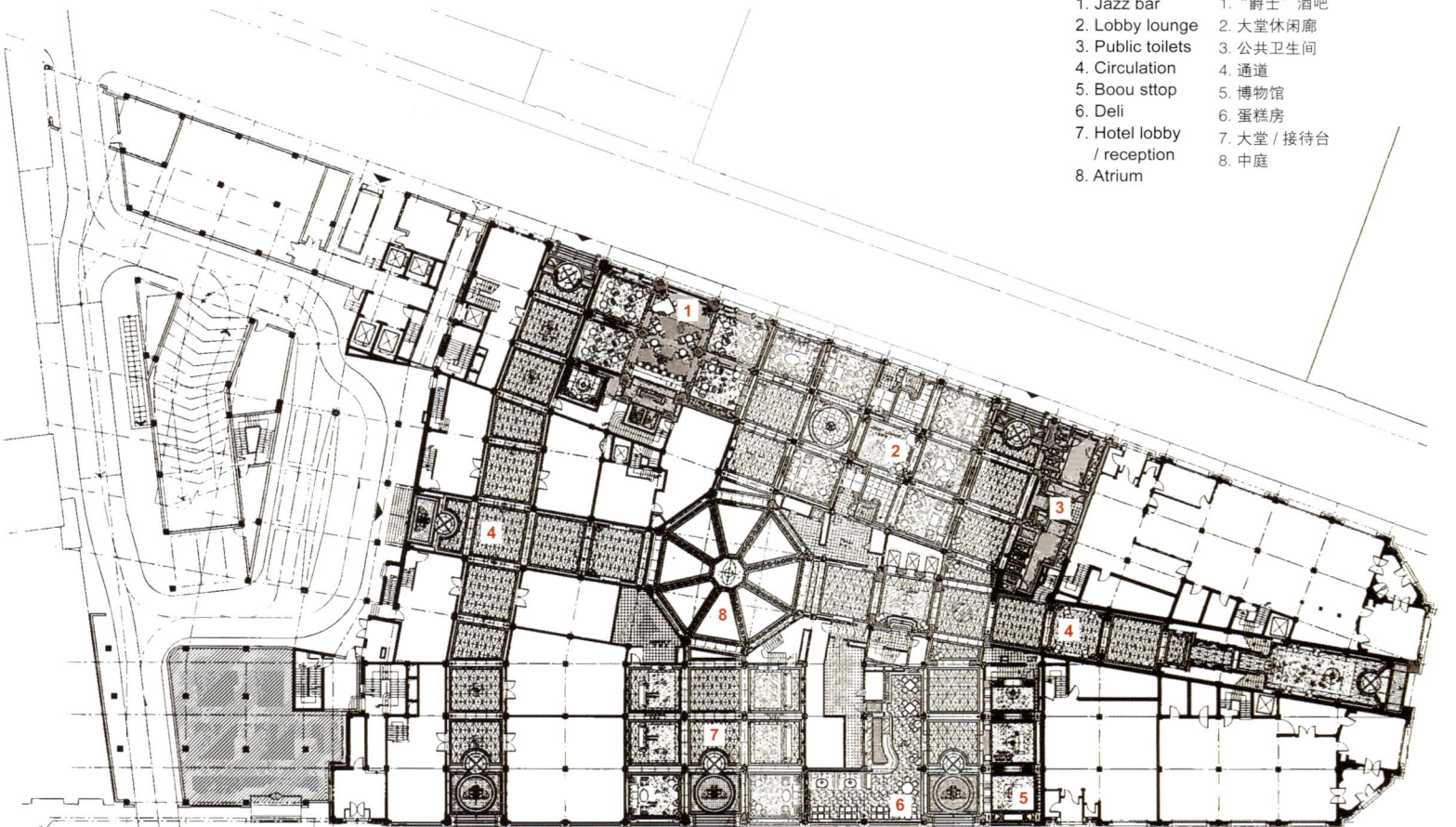

9

10

Nanjing Central Hotel

南京中心大酒店

Location: Nanjing, Jiangsu, China
Completion: 2010
Designer: W2 Architects
Design Team: Wang Degang, Yang Jing, Li Gan
Photographer: Li Gan

地点：中国 江苏省 南京市
翻新时间：2010年
设计：佳的建筑设计事务所有限公司
设计团队：王的刚、杨靖、李敢
摄影：李敢

BACKGROUND

The Nanjing Central Hotel is located in Xinjiekou - the bustling centre for business and commerce in the city, with convenient transportation and a lively market. The original structure of the hotel has attached an annex building in a circular shape with a diameter of 45 metres, which surrounds the semicircular main building. Due to limitations to the landscape, an expansion of the area is not possible.

SOLUTIONS

The opposite semi-circle was enclosed with glass, and transformed the above seven floors of exterior space into interior space. Through a simple and effective maneuver, the area of the hotel was increased by more than 1200 square metres, and a 25-metre-tall interior atrium was created. The originally less preferable views of the rooms on the interior circle has benefited from the introduction of the atrium, giving them a freshened outlook. A sightseeing elevator creates a vertical axis that transcends the main hall, which in practice fulfills transportation needs for 11 floors, but also corresponds to the aesthetics of the new enclosure.

In order for all the guests to enjoy the new view of the atrium, the front desk was moved from the first floor to the fourth floor and the buffet to the third floor, so that the best views are left for the guests. This organic restructuring of hotel functions allows practicality to marry with quality of service. The mastery of a unique style of architecture is paramount in allowing the building to stand out from other quality hotels in the surroundings. Instead of the simplistic, introverted mainstream style, the architects combined culture, aesthetics, market, and operation. Backed by an European style and adorned with modern techniques, the architects seek to create a new charisma for the hotel without losing its grace and classical charm.

After the renovation the composition of the hotel is more reasonable, and the utilisation of space is more complete, bringing the proprietor greater financial benefits. The perfect fusion of architecture, function, and decoration will bring a new view for the hotel.

1. Back elevation before renovation
2. Front elevation before renovation
3. Bedroom before renovation
4. Façade after renovation at night
1. 酒店翻新前背面外观
2. 酒店翻新前正面外观
3. 酒店翻新前客房
4. 夜色下的外观

5

背景

中心大酒店位于南京繁华商业中心——新街口，周围交通便利，商业发达。酒店原建筑的裙房平面为一直径约45米的圆形，客房主楼为外半环形，从而形成围合的内空间。由于用地及规划的限制，无法占用更多的地面面积。

过程

首先，设计师将圆形的另一边用通透玻璃幕墙围合，变原有四层以上室外空间为内部空间，以最为便捷有效的方法为酒店增加了1200多平方米的使用面积，且形成了25米高的内庭园中空大堂。原有景观最差的内圈客房，现在得以尽享中庭景观，使整个酒店形象焕然一新。配合围合的新空间而加设的观光电梯形成一条纵向的动态轴线，贯穿中空大堂上下，解决了从地面到11层的交通问题。

然后，总台区被从一层转换到了四层，自助餐厅改建到三层，让每个入住的宾客都能够最大限度的尽享中空大堂的景观，使各功能区有机结合，把窗外最好的风景留给客人，真正做到了功能与服务的完美结合。独特的设计风格，是其从周边众多高档酒店中脱颖而出的关键。设计师一改现今简约、内敛，强调低调奢华为主流的设计风格，把文化、美学、市场、营运完美结合。设计定位以欧式风格为背景，结合现代的装饰手法，以不失优雅和古典韵味为前提，让酒店焕发其自身的新魅力。

总之，重装后的整个酒店布局更合理，空间利用更充分，给业主带来更丰厚的经济效益。建筑、功能、装饰上的完美结合，让酒店呈现完美全新的形象。

5. Central atrium
6. Corner view in the atrium
7. Restaurant in the atrium

5. 中央心房
6. 心房一角
7. 心房内餐厅

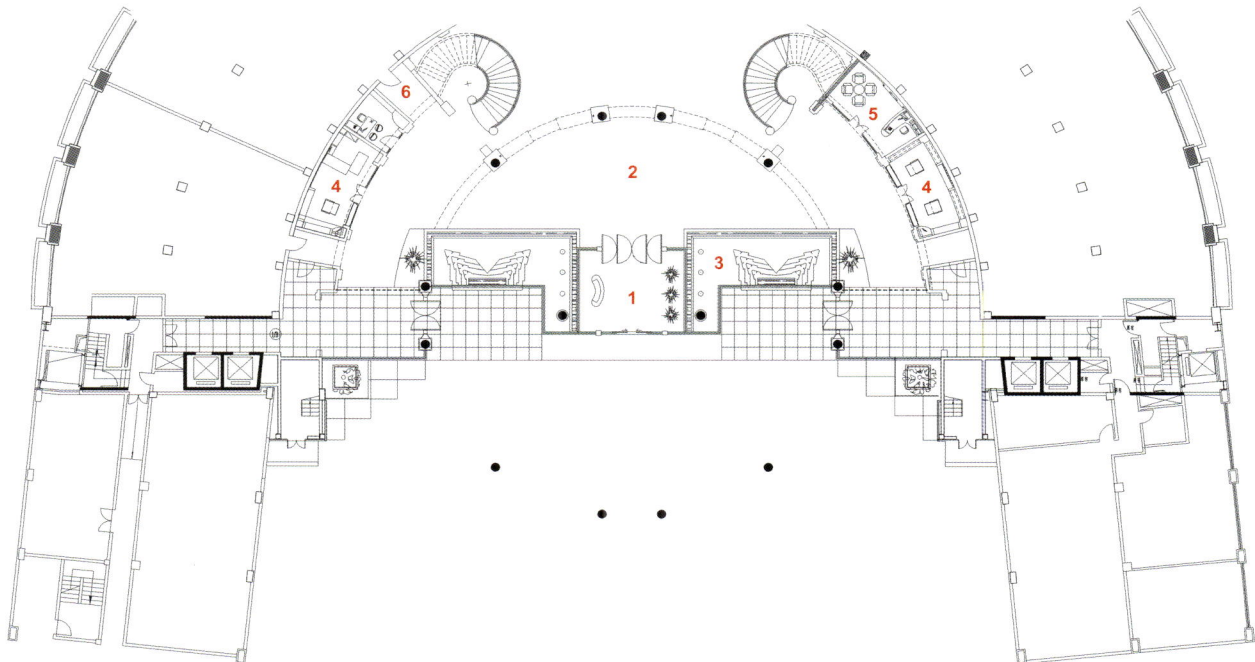

1. Entrance
2. Lobby
3. Fountain
4. Shop
5. Banquet reservation
6. Ticket centre

1. 入口
2. 大堂
3. 喷泉
4. 商店
5. 宴会区
6. 售票中心

8. Lobby on the ground floor
9. Restaurant on the second floor
10. Restaurant overlooking the streetscape

8. 一层大堂
9. 三层餐厅
10. 餐厅内可观看街景

9

10

13

11. Ceiling with patterns
12. Reception on the third floor
13. Coner view in the atrium
11. 天花板上的美丽装饰
12. 四层接待区
13. 中庭一角

Baronette-Renaissance Hotel

巴洛奈特文艺复兴酒店大堂翻新

Location: Novi, Michigan, USA
Completion: May 2010
Designer: David Ashen (d-ash design)
Photographer: Frank Oudeman
Area: 311m²

地点：美国 密歇根州 诺维市
翻新时间：2010年5月
设计：大卫·阿申（d-ash设计公司）
摄影：弗兰克·欧德曼
面积：311平方米

OBJECTIVE

The existing lobby was not that large, so the goal was to create an area that felt spacious and had a variety of experiences for the guest.

SOLUTIONS

Managed by Sage Hospitality, the Baronette Renaissance Hotel lounge designed by d-ash design is the leading representation of the Renaissance brand's new lifestyle directive.

Taking on the feel of a modern day lake retreat, a kind of oasis in Novi (the town in which the hotel is located), d-ash design's transformation of the public spaces of the hotel into the new Renaissance Lounge is inspired by American 20th century modernism. Therefore, they chose furniture pieces, colours and materials that were inspired by this era.

The hearth - an exceptional glass sculpture cast out of 30-inch-high glass - is the focal area of the grand lobby. Different areas for gathering radiate from the structure and create a number of opportunities for small groups or for individuals to work and reflect. Stone was reinterpreted as glass, creating a cascading and reflective wall. All materials are expressed true to their nature - steel, glass, wood working harmoniously together with the aim of redefining modernists' traditions.

Windows line either side of the lounge while the art, curated by Paige Powell, add dynamism to the space by featuring the strong creative energy that has emerged from the Detroit area during this not so bright time in this region's history. The artists chosen demonstrate that sometimes the most interesting artistic expressions emerge from the most difficult of times.

The overall feeling, therefore, was residential, with the focal point being the two-storey-high fireplace, around which designers placed comfortable and cosy seatings.

1. Reception before renovation
2. Rest area in the lobby before renovation
3. Area in front of the reception before renovation
4. White and black background wall coexisting and the grey, green, yellow colours boasting natural feeling

1. 酒店翻新前接待台
2. 酒店翻新前大堂休息区
3. 酒店翻新前接待台前方区域
4. 白色与黑色背景墙共存，灰色、绿色及黄色的运用突出了自然气息

5

目标

原有大堂空间较为狭小，因此设计的目标即为打造一个开敞的氛围，带给客人多样化的体验。

过程

酒店大堂是"新文艺复兴"理念的重要体现，设计师将其诠释为一个极具现代感的"湖畔居"这样一个概念。这一想法来自于美国20世纪现代主义风格，因此，家具、色彩及材质的选择上都注重突出这一时代风格的特色。

壁炉作为整个大堂的视觉中心，它的设计很特别，用了9米多高（大概是30英尺）特殊材质玻璃打造而成，整体造型突出了强烈的现代效果。不同的区域围绕着壁炉展开，为小规模的聚会等活动提供了完美的场所。玻璃取替石材，创造了一面似瀑布般的光滑墙壁。此外，所有的材质，如钢筋、玻璃、木材等都运用得恰到好处，相互融合再一次展现了"文艺复兴"的目标。

还有一点值得一提的便是窗户线。这是由设计助理佩奇·鲍威尔（Paige Powell）设计的，不但增加了整个空间的动感，还为室内带来很舒适的自然光，映射出底特律地区在历史上最为艰难时期中爆发出的强大而又别具创意的能量。

大堂内整体呈现出居家氛围，设计师更在壁炉的周围摆放了舒适的座椅。

5. Reception desk
6. The Khaki-coloured leather chairs being extremely attractive in the natural-coloured space
7. Distinct decorative objects on the wall
8. Specially designed chairs in the lounge

5. 接待台
6. 土黄色的皮面座椅在中性色调为主的空间中格外引人注目
7. 墙壁上别具特色的装饰品
8. 休息区内设计独特的座椅

9. Reception desk on the one side and communal table on the other side with cute lighting fixtures dangling from the ceiling
10. Pure white shelves showcasing various snacks serving as decoration objects
9. 接待台设置在一侧，另一侧摆放着长桌，可爱的灯饰从天花上悬垂下来
10. 白色的架子上摆放着各种零食，犹如装饰品一般

10

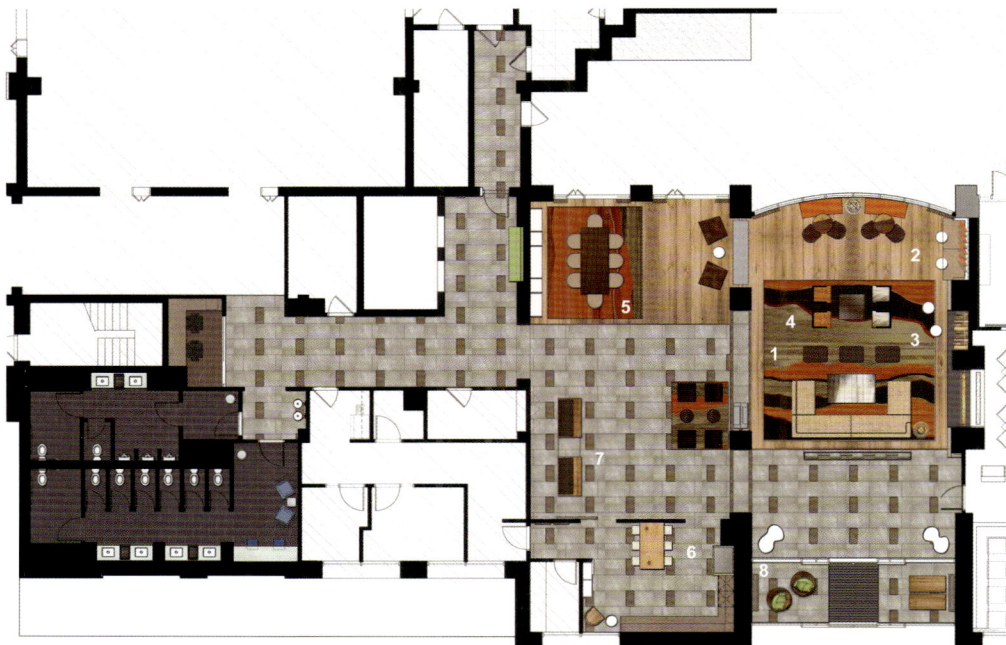

1. Lobby
2. Lobby Lounge
3. Hearth
4. Lobby detail
5. Public work space
6. Public space
7. Reception
8. Lounge area
1. 大堂
2. 大堂休息区
3. 壁炉区
4. 大堂细节设计
5. 公共办公区
6. 公共区
7. 接待台
8. 休息区

Swissotel Duesseldorf / Neuss

杜塞尔多夫瑞士酒店

Location: Duesseldorf, Germany
Completion: 2009
Designer: Brandherm+Krumrey Innenarchitektur in
          collaboration with Ina Rinderknecht
Photographer: Joachim Grothus
Area: 15,000m²
地点：德国 杜塞尔多夫
翻新时间：2009年
设计：Brandherm+Krumrey 室内设计公司与 Ina Rinderknecht 合作
摄影：乔基姆·格罗休斯
面积：15000平方米

OBJECTIVE

Brandherm+Krumrey, Cologne, was in charge of designing and planning the conversion of all of the areas of the business hotels: the room floors, the lobby and reception area, the restaurants, fitness and spa area, corridors and all of the conference facilities. The entire conversion was carried out in four building phases. Another major task was to integrate the installations required under strict fire-protection codes invisibly into the overall structure.

SOLUTIONS

The formal design is supplemented by materials of high quality such as woods, richly varied natural stones, glass elements and metallic surfaces. The extraordinary combination of this materiality is also reflected in the selection of fabrics, with a wide variety of textures and nuances of colour fine-tuned down to the last detail. The carpet designs emphasise the overall concept. Straight away, the beholder sees that these designs, while lively, still fit in with the simple styling of the overall design.

The guest experiences the new, consistent and authentic character of the Swissotel immediately upon entering the lobby. The rectilinear reception desk enveloped in marble is reflective of this urbane expression. A direct relationship exists between this and the flowing transition to the bar with lounge area.

Planned down to the details, this interior design permits a relaxing feeling of well-being. Individual zones in the lounge, for lunching or simply for enjoying, have a positive effect. Easy chairs and high 3-seaters, niche designs offer the guest large-area communication but also the prospects of privacy as well. The accents of the lighting concept are intentionally individually placed as floor lamps in combination with suspended fixtures in various sizes bathe the entire area in warm light.

An assortment of scenarios is possible in the hotel's conference areas. The flexibility, both in terms of room size but also seating (with or without tables), a variety of seating arrangements, or different orientation options, make them well suited for use as conference and event areas. Within the conference areas, an assortment of light scenarios can be called up, e.g. the cosy candlelight dinner, spot lighting during lectures, or complete lighting for trade-fair events. The use of multifunctional technologies featuring the latest developments and yet practically invisible, rounds out the picture.

A particular challenge was to put a new face on the "Jupiter" ballroom, an area totalling 1,500 square metres, while at the same time doing justice to the recognition factor and the variety of uses to which the ballroom is put.

With clear, closed geometric shapes, the uniform concept is communicated directly to the guest. Stylish, timeless and with a simple elegance, in its new garb the Swissotel makes a generous impression indeed.

1. Lobby before renovation
2. Restaurant before renovation
3. Conference room before renovation
4. Entrance
5. Marble reception desk
1. 酒店翻新前大堂
2. 酒店翻新前餐厅
3. 酒店翻新前会议室
4. 入口
5. 大理石接待台

6

目标

Brandherm+Krumrey室内公司负责杜塞尔多夫瑞士商务酒店所有区域的翻新工作，包括客房、大堂、接待台、餐厅、健身馆、水疗区、走廊以及会议室。整个翻新工程分为四阶段完成，另一重要任务为在所有的结构内安装可见防火代码装置。

过程

高品质的材质，如木材、天然石材、玻璃及金属等更完善了酒店内庄严典雅的氛围，而各种材质的完美搭配更体现在制品的选择上，多样的纹理花样以及变换的色彩深入到每一处细节中。其中，地毯设计便是很好的诠释，高雅却不失与整体的简约风格相协调。

客人从步入大堂开始便可体会到新鲜的、连续而真实的特色。直线型的接待台"包裹在"大理石材质中，突显出都市气息；休息区将接待处及酒吧"衔接"，展现出连续统一的风格。

此外，设计中注重营造轻松愉悦的氛围，休息区内的单独座区为聚餐或休憩提供了完美的场所。安乐椅以及三人座椅更是为客人提供了交流的空间，开敞而不是私密性。灯饰特意单独装置，落地台灯与规格不一的吊灯使得整个大堂沐浴在温和的光线中。

酒店会议区设计需要注重灵活性，不仅体现在房间大小，更体现在座区排列及朝向上，以便于适用于不同类型的会议及各种活动。同样，会议区内灯光的选择也要注意多样性，如烛光晚餐、聚光灯演讲或者全照明交易会展。高新科技的运用可以更好地解决这一问题。

设计面临的一个特殊挑战即为如何为原来的"木星舞厅"（面积达1500平方米）打造全新形象，同时不失功能性。

总之，全新的设计以其清晰而封闭的几何造型为客人带来耳目一新的体验。时尚、永恒与雅致在这里共同呈现。

6. Restaurant
7. Bar area
8. Lounge
9. Conference room
10. Meeting room
11. Gym and pool
6. 餐厅
7. 酒吧区
8. 休息区
9. 会议室
10. 小会议室
11. 健身房及泳池

7

8

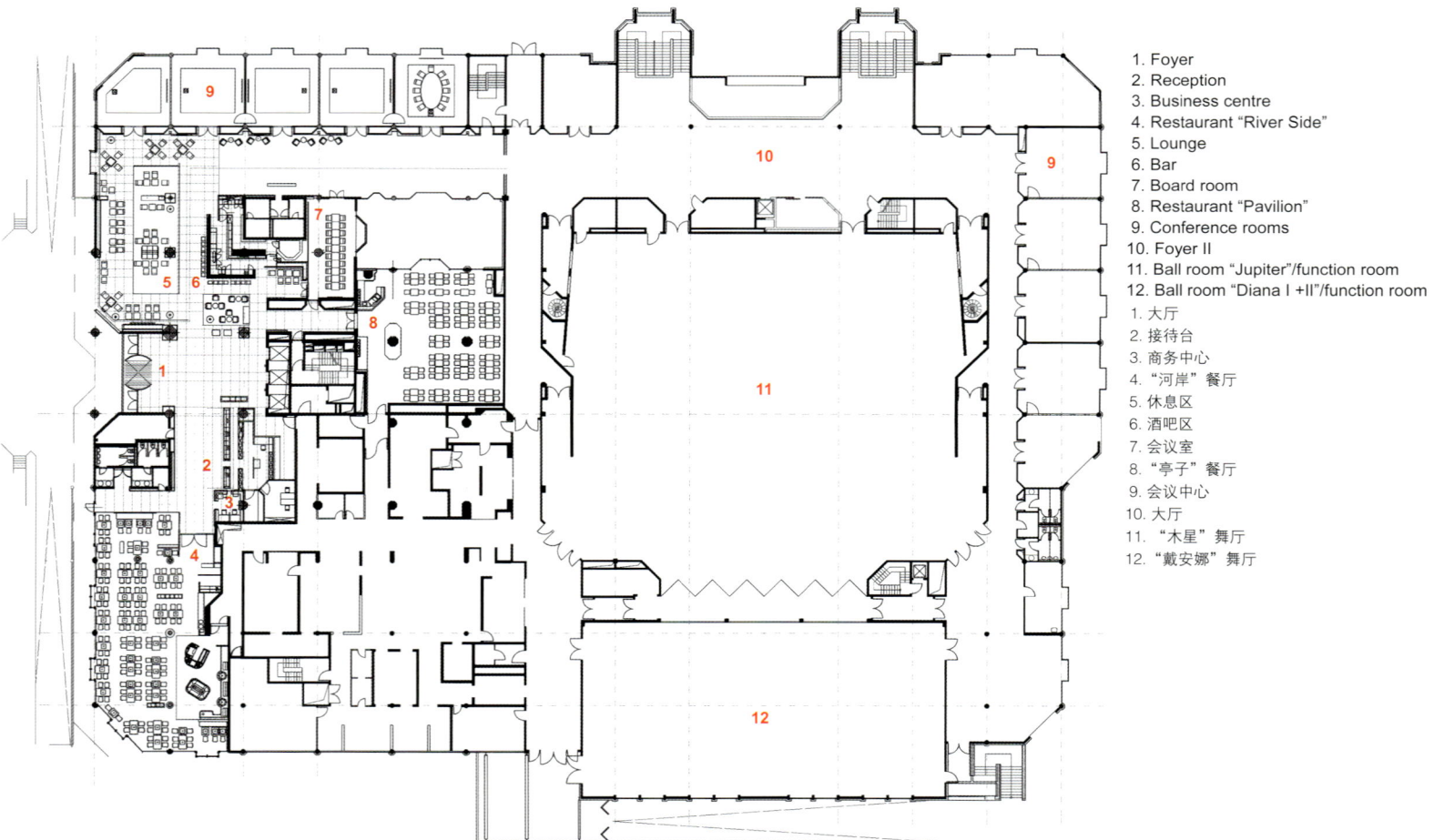

1. Foyer
2. Reception
3. Business centre
4. Restaurant "River Side"
5. Lounge
6. Bar
7. Board room
8. Restaurant "Pavilion"
9. Conference rooms
10. Foyer II
11. Ball room "Jupiter"/function room
12. Ball room "Diana I +II"/function room

1. 大厅
2. 接待台
3. 商务中心
4. "河岸"餐厅
5. 休息区
6. 酒吧区
7. 会议室
8. "亭子"餐厅
9. 会议中心
10. 大厅
11. "木星"舞厅
12. "戴安娜"舞厅

12

12. Corridor
13. Living room in the suite
14. Bedroom
12. 走廊
13. 套房起居室
14. 卧室

Hotel Principe di Savoia

萨沃尔王子酒店

Location: Milan, Italy
Completion: 2009
Designer: CDA Design, Francesca Basu Designs Limited,
Thierry W Despont Office
Photographer: Guillaume de Laubier

地点：意大利 米兰
翻新时间：2009年
设计：CDA 设计、弗朗塞斯卡·巴苏设计有限公司、蒂埃里·W·德蓬公司
摄影：威廉·德·罗毕尔

OBJECTIVE

Hotel Principe di Savoia, one of the Dorchester Collection has undergone an exciting transformation. While the lobby, Il Salotto, and bar were redesigned by acclaimed international architect, Thierry Despont; nine new Principe suites have also been introduced and the Imperial Suite has undergone a complete transformation. London-based architect Francesca Basu is responsible for the restyling of these nine new Principe suites following her sensational design of the mosaic suites last year. Thierry's vision is also to bring the Principe Bar to life, making it more vibrant and engaging, while retaining some of the original features such as the beautiful marble and wood panelling.

SOLUTIONS

The lobby, Il Salotto, is now a much lighter, more welcoming entrance to the palace hotel. On arrival, the promenade lounge on the right hand side is a convenient meeting space for the well-heeled Milanese and guests alike to enjoy an aperitif surrounded by sumptuous Italian furniture. To create an immediate impact on arriving at the Principe di Savoia, Despont has added classical paintings by famous artists such as Luca Giordano and a spectacular custom-made Murano glass chandelier created by Barovier and Toso. This total transformation creates the appropriate feeling of grandeur and occasion for those arriving at this iconic hotel.

Despont has managed to create a perfect balance between the classical and innovative styles that embody the Principe di Savoia. The centrepiece of the room is to be a custom-made banquette that will "wrap around" a grand piano, perfect for intimate musical soirees.

The redesign of the hotel's famous bar further enhanced its legendary sumptuous style. The bar is a work of art in itself; sculpted tinted crystal with a back-lit mirrored wall to perfectly complement the light playing round the room. Thierry uses words such as "light, sparkle, life, sensuous, glittering and radiant" to describe his newly designed bar, which will be the most talked about re-vamp in Milan.

The Principe suites are appointed with hand-painted frescoes, traditional Italian furniture as well as deep purple armchairs and floor-lamps creating a warm ambience. The large sitting rooms, part of each suite, are furnished with sumptuous sofas while colourful Murano glass vases sit on Lombardy style tables. The stunning bathrooms are spacious, light and equipped with a Lasa marble bath in the centre of the room as well as a shower adorned with striking hand made glass mosaics.

The redesign of the Imperial Suite was overseen by Celeste dell'Anna and combines a series of contemporary and classical elements. The 230-square-metre suite has a large sitting room, featuring a handcrafted mini bar console and crocodile skin writing desk. Striking paintings featuring interpretations of contemporary masterpieces have been specially created by Celeste dell'Anna. The bedroom boasts a four-poster bed in the richest fabrics and a large walk-in wardrobe capable of housing all the latest purchases from the nearby shopping district. One of the highlights of this suite is the Turkish bath which, with its chrome-therapy system, massaging shower and special make up console, encourages complete relaxation, a luxury that is taken very seriously at the Hotel Principe di Savoia.

1. Sitting room in Presidential Suite before renovation
2. Living room before renovation
3. Suite before renovation
4. Entrance
1. 翻新前总统套房起居室
2. 酒店翻新前套房客厅
3. 酒店翻新前套房
4. 入口

5

目标

萨沃尔王子酒店隶属于多切斯特酒店集团，经历了一次大规模的翻新。大厅、酒吧等公共区域的改造由国际知名建筑师蒂埃里·德篷（Thierry Despont）负责，而9间新套房的风格则由弗朗塞斯卡·巴苏（Francesca Basu）全面设计。酒吧区的设计目标即为使其更加活力十足，但同时保留原有的特色，如大理石及木板等结构。

过程

大厅更加开阔明亮，洋溢着友好的氛围；休息区位于入口右侧，为米兰上层人士及酒店客人提供了一个方便的聚会场所，更可以在这里小饮几杯开胃酒，同时欣赏周围摆放着的意式奢华家具。为给到来的客人留下深刻的印象，设计师特意悬挂了几幅出自艺术大家的画作以及专门定制的穆拉诺玻璃吊灯（Murano：意大利知名玻璃饰品品牌）。整体的改变让那些前来酒店的客人感受到了舒适而不失华贵的气息。

设计师尽量在古典与创新中取得平衡，大堂中最为吸引眼球的当属摆放在钢琴四周的定制长条形软座椅，为一场黄昏音乐会提供了完美的场所。

套房增添了手绘壁画、古典意式风格家具、深紫色扶手椅以及落地吊灯，温馨祥和的氛围应运而生。客厅变得更加开敞，豪华的沙发以及摆放在桌子上的穆拉诺彩色玻璃花瓶格外吸引眼球；浴室宽敞明亮，大理石浴缸摆放在中央，淋浴器采用手工玻璃马赛克装饰。

帝王套房的重新设计由室内设计师赛莱斯特·德尔·安娜（Celeste dell'Anna）负责，现代元素与古典元素在这里共存。套房面积为230平方米，带有宽大的客厅，手工制作的迷你吧台以及鳄鱼皮面书桌构成主要特色，现代风格的画作全部由设计师亲自创作。卧室中的四柱床格外引人注目，宽敞的步入式衣柜可以容纳客人购买的各种物品。这里最具特色的当属土耳其风格浴室，带有光谱色疗系统、按摩淋浴器、化妆台等，让人尽情的放松。

5. Lobby lounge
6. Principe Bar
7. Counter and cellarette in Principe Bar
5. 大堂休息室
6. "王子"酒吧
7. "王子"酒吧吧台及酒柜

8

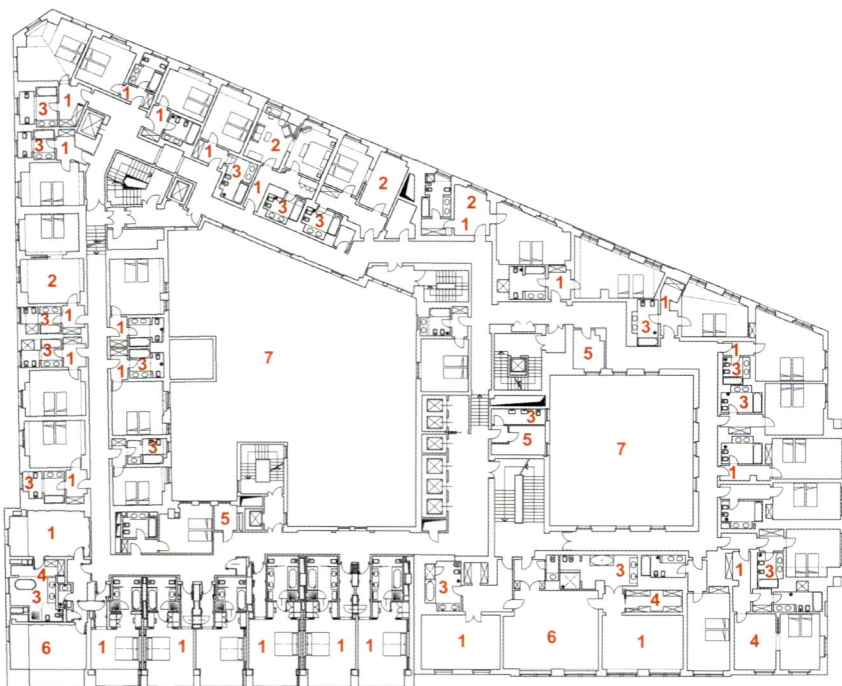

1. Room
2. Living room
3. Bathroom
4. Dressing room
5. Office
6. Lounge
7. Courtyard

1. 客房
2. 起居室
3. 浴室
4. 化妆间
5. 办公区
6. 休息区
7. 庭院

8. Living room of the Imperial Suite - red candles, ceramic vases, round lampshades as well as framed mirror bringing noble and luxurious feeling to the whole atmosphere
9. Living room of the Imperial Suite - the framed paintings hung on the wall adding a touch of interest in the relatively sumptuous space
8. 帝王套房客厅内的红烛、陶瓷花瓶、圆形灯罩以及镶框镜子等装饰品营造了高贵奢华的气息
9. 帝王套房客厅内墙壁上悬挂着的画框为这个华丽的空间增添了一丝趣味性,别具特色

10

10. Bedroom of the Imperial Suite - colours of white, red and cream dominating every corner
11. Bedroom of the Principe suite - distinctively framed mirror and painting on the wall being the main features
12. Living room of the Principe suite - red being the main tone

10. 帝王套房卧室内，白色、米色及红色构成了主要基调
11. 王子套房卧室内，形状独特的镜子以及墙壁上的画框格外吸引眼球
12. 王子套房客厅内，红色成为了主要色调

Sofitel Lyon Bellecour

索菲特酒店

Location: Lyon, France
Completion: 2009
Designer: Studio Patrick Norguet
Photographer: Renaud Callebaut
地点：法国 里昂
翻新时间：2009年
设计：帕特里克·诺尔盖工作室
摄影：雷诺·嘉勒博

OBJECTIVE

Built in the 1970s and the backdrop to summits and top-level meetings in the past, its delightfully harmonious contrasts, the unexpected authenticity of its contemporary interior and the familiar modernity of its scale are what give this hotel a whole new appeal. The aim is to give the hotel a true sense of identity and to avoid the merely decorative. The designer's "slow" approach to restyling the hotel's interior is behind this unique feel, giving it that timeless quality that will stand the test of time. For "slow design" is not simply a passing trend, but a whole new concept, an alternative approach to design. It represents a completely new lifestyle with optimum use of available resources and a life led in harmony with the surrounding environment.

SOLUTIONS

The design of the common areas of the hotel, including the lobby, the Le Melhor bar, the Silk brasserie, the Les Trois Dômes gourmet restaurant, the gym, the spa and the adjoining garden with its bar, is the result of an extremely detailed contextual analysis by the designer. Patrick Norguet delved ever deeper into Lyon's wealthy past as nerve centre of the French textile industry, and of the silk industry in particular. Wishing to pay homage to an evocative national cultural heritage, Norguet joined forces with Tassinari et Chatel, one of the city's oldest and the last remaining textile merchants. Together they performed the painstaking task of trawling through historical records, reproducing each and every motif in existence over the years. The result is a patchwork of embroidery and prints that decorates the walls of the lobby, in perfect contrast with the contemporary décor. The designer called on the services of artist Gille Cenazandotti for this part of the project.

The sustainable and convivial sensory aspects of Norguet's new interiors also feature harmonious contrasts in the décor, such as the coffee and silvery colours in the Trois Dôme restaurant and the red and black in the Le Melhor bar, the whole theme punctuated by understated yet original furnishings with attitude. The convivial, narrative style of the design whisks us back, quite effortlessly, to the bosom of the family, a nurturing cocoon where stories play out, relationships forged, where there is room for meetings and exchanges, developing and creating, as if by magic, new cyclical patterns.

This recent renovation, has succeeded in bringing to the fore and the prestigious character of a hotel that is a landmark in Lyon. The new sleek, yet welcoming and convivial interior is the product of the perfect complementary mix of traditional craft techniques and the designer's eclectic choice of the very finest of materials.

1. Lobby before renovation
2. Reception before renovation
3. Restaurant before renovation
4. Lobby lounge
1. 酒店翻新前大堂
2. 酒店翻新前接待台
3. 酒店翻新前餐厅
4. 大堂休息区

目标

原建筑始建于20世纪70年代，这里曾是各种峰会及重要会议的举办地，现代风格的室内空间以及规模将赋予其全新的面貌。设计的目标即采用"慢设计"理念，打造永恒的品质，使其经得起时间的考验，同时体现酒店的特色而避免单纯的修饰。"慢设计"并不是一种转瞬即逝的流行趋势，而是一种全新的理念，一种独特的设计方式。它更代表着一种全新的生活方式——合理利用资源，与周围环境和谐共处。

过程

设计师在对大堂、酒吧、餐馆（以售啤酒为主）、美食餐厅、健身房、水疗馆以及花园等公共区域进行翻新设计之前，对其原有状况进行全面的分析。设计中更深入了解里昂纺织业的历史，旨在酒店中体现这一文化遗产的重要地位。诺尔盖与最古老的纺织商合作，查阅纺织业历史，重现了每个时期不同的图案样式，最终打造了带有精美刺绣图案的花布，用于装饰大厅的墙壁，与整体的现代风格装饰形成完美的对比。

设计中注重营造永恒感及愉悦的空间氛围，这同样体现在装饰对比上，如餐厅中咖啡色及银色的对比、酒吧中红与黑的对比。此外，欢快的氛围让人不禁想到与亲朋好友的相聚，这里设计有会议室及交流区，满足了客人的需求。

酒店翻新工作由法国设计师帕特里克·诺尔盖负责，成功将酒店特色展现出来。时尚、友好而喜悦的空间氛围在很大程度上取决于传统纺织技艺与设计师精选材质的完美结合。

5. Fireplace in the lounge being extremely eye-catching as well as creating warm atmosphere
6. Specially-shaped chairs in the lounge
7. Restaurant - the ceiling and wall being decorated with sculpture patterns
8. Restaurant
9. Bar area - the purple chairs giving off elegance while the white counter highlighting simplicity and brightness

5. 休息区内的壁炉营造了温馨的氛围，格外吸引眼球
6. 休息区内造型独特的座椅
7. 餐厅区内，屋顶及墙壁上装饰着雕塑图案
8. 餐厅
9. 酒吧区内，紫色座椅令另空间更加雅致，白色的吧台则格外明亮并突出简洁的设计理念

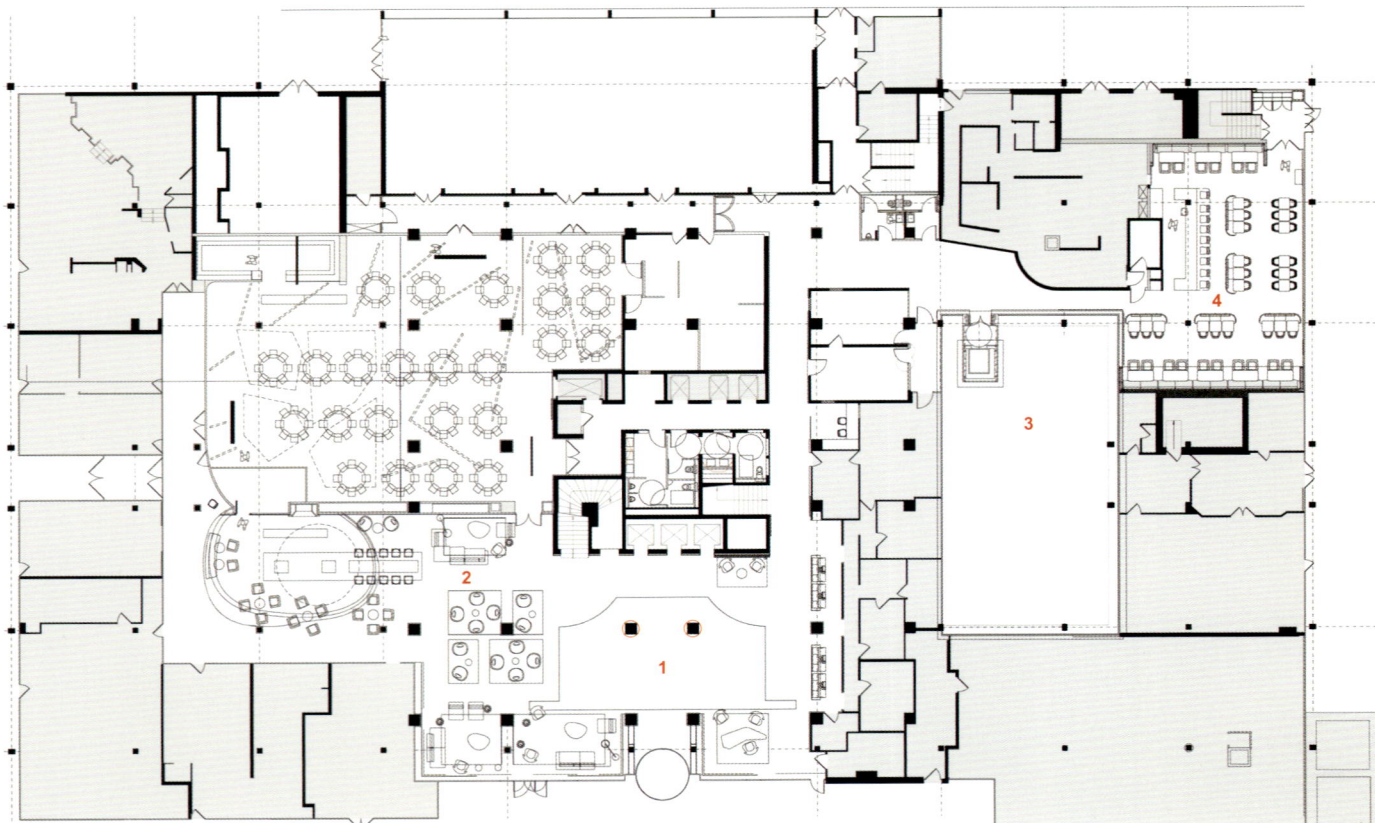

1. Lobby
2. Lobby bar
3. Fitness/spa garden
4. Silk Brasserie
1. 大堂
2. 大堂酒吧
3. 健身中心/水疗馆
4. 餐厅

Hotel Monaco Alexandria

亚力山德里亚摩纳哥酒店

Location: Alexandria, Virginia, USA
Completion: 2008
Designer: Cheryl Rowley Design, Inc.
Photographer: Cheryl Rowley Design, Inc.
Project Budget：$22 million
地点：美国 弗吉尼亚州 亚力山德里亚市
翻新时间：2008年
设计：谢丽尔·罗利设计事务所
摄影：谢丽尔·罗利设计事务所
预算：2200万美元

BACKGROUND

On the West banks of the Potomac River lies the colonial town of Alexandria, overlooking what was once a bustling harbour - a historic port of entry for vessels traversing the high seas engaged in international and coast-wide trade. Having served several times as a civil war battleground, as well as being home to a young mapmaker named George Washington, the city is deeply etched in American history.It is here that the Hotel Monaco Alexandria makes its home at 480 King Street in the historic district of Old Town.

SOLUTIONS

In its approach to renovating and rebranding this former Holiday Inn Select, the design team began with the signature "Hotel Monaco" brand design elements to create a feeling of comfort and welcome, but also an expectation of discovery, novelty and whimsy. Taking its inspiration from the city's rich military history and a culture of sea-faring travel, the hotel's interior is an ornately rich environment that echoes Alexandria's history and fuses it with the adventurous forward-moving spirit of exploration abroad.

A lively interior beckons passersby to peek through the street front windows. The traditional wood panelled walls are painted a vibrant "Naples Blue" - emulating the strong tonal colours of early colonial interiors.The ebony chevron patterned wood floor offers a sharp contrast to the bold walls and completes the saturated architectural backdrop for this eclectic space. The Monaco signature "trunk" design of the registration desk is given a twist by way of cardinal red leather covering and exquisite nail head detailing. Just behind the desk stand meticulously stenciled panels covered in an arabesque pattern. An oval dome in the lobby's ceiling provides a natural focal

point and is punctuated by a dynamic chandelier composed of multicoloured glass and silk orbs. Beyond this central piece lies an inviting fireplace. A faux painted leopard skin pattern adds exotic flavour to the classically styled console, an inviting locale for a rich glass of wine and friendly conversation.

In the living room area of the lobby, exaggerated classical furniture silhouettes are accentuated by richly textured fabrics vivid in colour and patterns. The warm palette of dusty reds, Asiatic blues, rich browns, and soft beiges, entices the guest to relax in one of various seating arrangements.Set under the lobby's dome, hand-tufted area rugs emphasise the Living Room with their richly coloured graphic patterns and Moroccan flavour.

The window treatments are Roman Shades of a sheer cream fabric with black edge banding and tassled pulls.These romantic shades filter the sun, but also leave an open and inviting view of the space for passersby.Hand painted Moroccan antiques scattered throughout the space emanate the splendor of far away lands.

1. Lobby before renovation
2. Corridor along the guestrooms before renovation
3. Guestroom before renovation
4. Lobby
1. 酒店翻新前大堂
2. 酒店翻新前客房区走廊
3. 酒店翻新前客房
4. 大堂

5. Lounge area
6. Reception desk
7. Corridor along the guestrooms with a totally new look
8. Living room of the guestroom - chair and carpet with the same pattern highlighting each other
9. Living room of the guestroom - red chairs making the space more warm

5. 休息区
6. 接待台
7. 客房区走廊焕然一新
8. 客房起居室内，座椅与地毯带有相同的图案，相互呼应
9. 客房起居室内，红色的座椅使得空间更加温馨

背景

亚力山德里亚这一殖民小镇坐落于波托马克（Potomac River）河畔。这里曾经是一个繁华的国际贸易港口，曾多次作为内战战场，更是乔治·华盛顿（George Washington）的故乡，在美国历史上具有深远的影响。摩纳哥酒店正是坐落在这一小镇的著名历史区内——480国王大街。

过程

设计团队肩负两项任务——酒店空间翻新及品牌重塑（原酒店名为"假日精品之家"）。他们将品牌重塑作为出发点，充分运用"摩纳哥酒店"设计元素，营造一个舒适、友好的氛围，同时突出创新与特色，为客人打造独特体验。室内翻新从酒店所处地理位置的丰富军事历史及航海旅行中获得灵感，华丽而多彩的空间彰显出亚力山德里亚的历史背景的同时，更体现出勇于向前的探索精神。

室内风格活泼，吸引着路人透过沿街的窗户不断向内眺望。古老特色的木板墙饰以"那不勒斯"蓝，以模仿早期的殖民地风格；黑檀木材质"山字形"地板与墙壁形成强烈对比，同时为中性风格空间带来独特的设计元素；"摩纳哥酒店"标志性元素——象鼻型接待台被赋予了新的特色，采用鲜红皮革饰面并运用指甲图案装饰；接待台后面"矗立"着交错排列的装饰板；大堂天花上椭圆形的拱顶自然成为了中心，多色玻璃及丝绒球组成的玻璃吊灯更增添了动感；温暖的壁炉异常引人注目；人造豹纹图案带来异域气息。

大堂休息区内，造型夸张的家具在纹理丰富、色彩及图案多样的织品的映衬下更加突出。座区的颜色格外丰富，暗红色、蓝色、褐色、米色应有尽有，吸引着客人坐下来小憩。拱顶的下方，手工缝制的小毯子以及色彩丰富的几何图案、摩洛哥风格共同强调着浓郁的家居氛围。

乳色的罗马窗帘不仅可以过滤阳光也可以让人清晰地瞥见内部的景象，手工喷漆的摩洛哥古玩随意摆放着，呈现出独特的韵味。

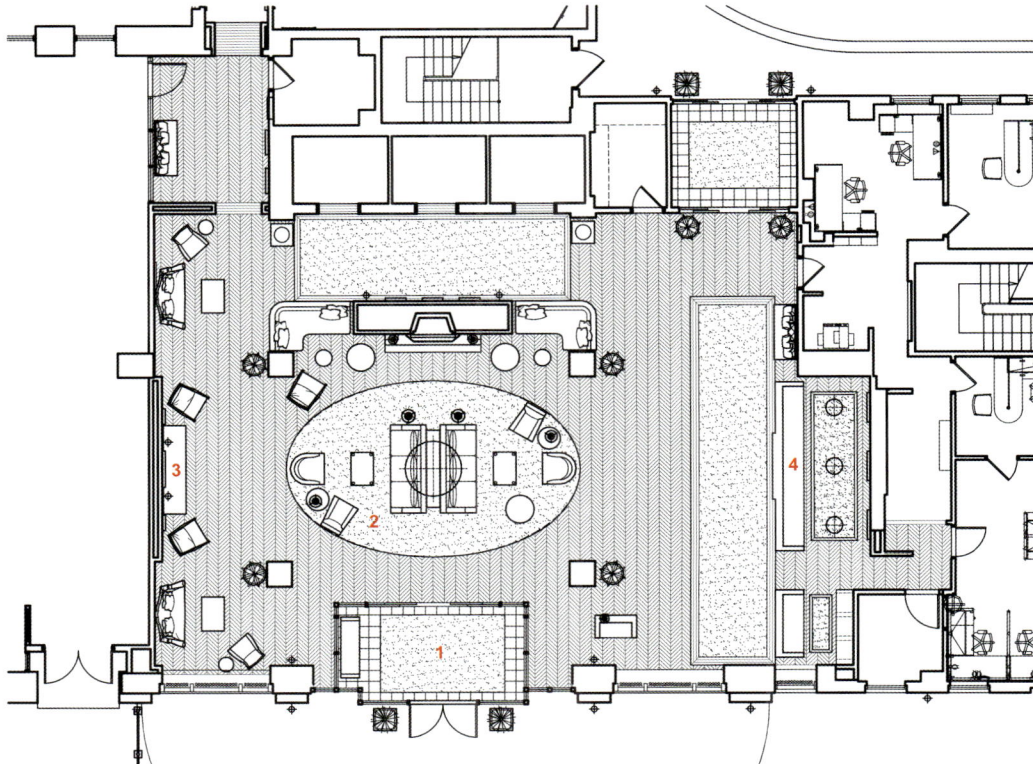

1. Foyer
2. Lobby
3. Lounge
4. Reception
1. 入口大厅
2. 大堂
3. 休息区
4. 接待台

10. Guestroom
11. Bathroom of the guestroom - the mosaic patterns with only
    black and white colours boasting simplicity and cleanness
10. 客房
11. 客房浴室墙壁上黑白两色马赛克图案设计凸显简约与简洁

Hotel Modera

墨德拉酒店

Location: Portland, Oregen, USA
Completion: June 2008
Designer: Corso Staicoff, Inc.
Photographer: Jeremy Bittermann, David Phelps, Dan Tyrpak
Area: 11,098m²
地点：美国 俄勒冈州 波特兰市
翻新时间：2008年6月
设计：科尔索·斯泰科夫设计公司
摄影：杰里米·比特曼、大卫·菲尔普斯、丹·泰尔帕克
面积：11098平方米

OBJECTIVE

The Hotel Modera is a complete renovation - transforming a neglected motor lodge, often thought of as an eyesore, into one of Portland's sleekest places to stay. This new design restores the simple lines of the 1964 building and pays homage to the mid-century origins.

SOLUTIONS

This project was a big challenge due to a conservative budget and fast-track schedule - the design and construction was completed in a 10-month period. The design team had to quickly decide where it was appropriate to focus the dollars in order to obtain the maximum result. Each guest floor plan (aside from six suites) remain as originally constructed, allowing more opportunities to create a dramatic lobby and courtyard.

The interior team worked with the architects and landscape architects to reconfigure the old building and its entry sequencing. The porte cochiere was relocated, the lobby was extended and a central courtyard was added, replacing the former parking lot. The new lobby's floor-to-ceiling glass connects the courtyard and integrates the indoors with the exterior. Walnut flooring that wraps to the walls, together with Calcutta marble and rich fabrics, meld the simplicity of the architecture with warm tones and inviting textures as well, allowing sculptural furnishings and provocative artwork to steal the show.

The guestrooms - with a minimal colour palette comprised of rich browns, reds and orange - are arranged to maximise small spaces for function and comfort. Crisp white bedding, flat screen televisions, and iPod clock radios add to the luxurious feel and provide the discerning traveller with the comforts of home.

1. Exterior view before renovation
2. Lobby before renovation
3. Guestroom before renovation
4. Façade with a new look
1. 酒店翻新前外观
2. 酒店翻新前大堂
3. 酒店翻新前客房
4. 翻新后外观

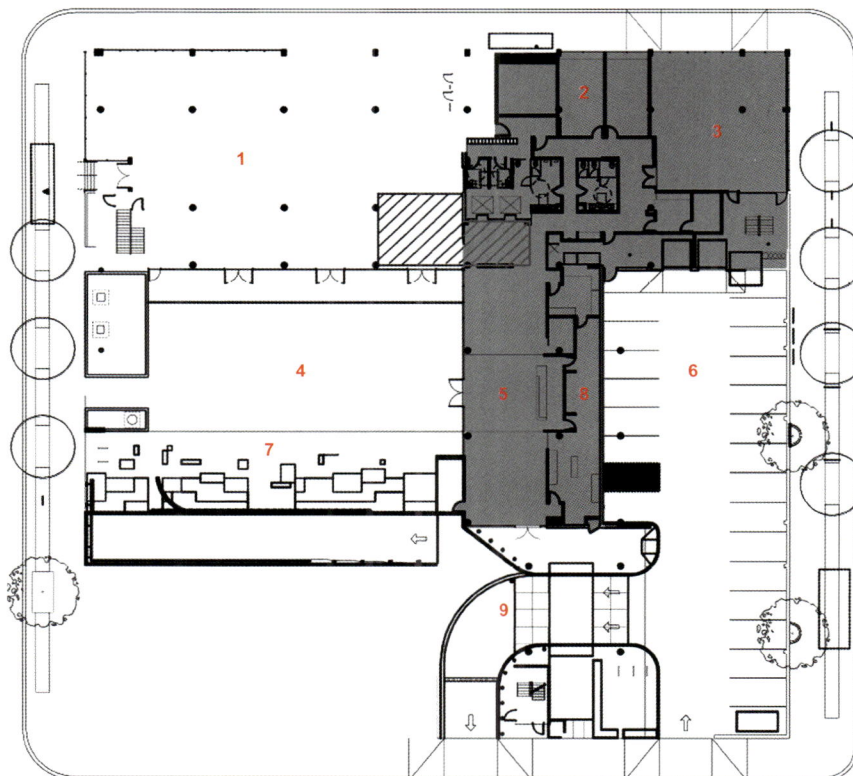

1. Restaurant
2. Meetingroom
3. Ballroom
4. Courtyard (former parking)
5. Lobby
6. Parking
7. Living wall & green roof
8. Staff
9. Porte Cochire

1. 餐厅
2. 会议厅
3. 舞厅
4. 庭院（原停车场）
5. 大堂
6. 停车场
7. 栽种植物的墙壁和屋顶
8. 员工入口
9. 可让车辆出入的庭院

5

目标

墨德拉酒店是一个全面翻新的项目，是将原来一处很少有人光顾的汽车旅馆——一处被人看作碍眼的地方改造成波特兰市一座最迷人的酒店。新的设计再现了这座1964年建成的建筑的简洁线条，并忠实于中世纪的传统。

过程

这个项目对设计师来说是个挑战，因为项目的资金有限，时间紧——设计和建造在10个月内完成。设计小组必须很快决定如何将钱花在刀刃上，以取得最好的效果。每层客房的平面布置（除了6间套房以外）都保持原样，而将重点放在建造一个辉煌的大堂和庭院。

室内设计小组与建筑和景观设计师重新设计了建筑及其入口的次序。车辆出入庭院的通道被重新布置，大堂被延伸；并加建了一个中央的庭院，替代了原有的停车场。新大堂的落地玻璃窗系室内外，与中央庭院视野相连。大堂内，核桃木地板、核桃木墙面，连同加尔各答产的理石、厚实的装饰织物与构造简洁的线条相融，增添了酒店温暖氛围和亲切的特征。而如雕塑一般的家具和古味盎然的艺术品也在这里非常抢眼。

客房的色彩简洁，用深棕色、红色、橘黄三色使不大的空间最大化、功能化，给人以舒适之感。清新的白色床单、平板电视、ipod时钟收音机在给客人一种奢华的感受的同时，还给敏感的游客家一般的舒适。

5. Reception desk with exquisite chandelier dangling from the ceiling
6. Chairs of different materials and shapes furnishing the lounge
7. Guestroom
5. 接待台上方精美的吊灯从天花上悬垂下来
6. 休息区内材质及形状各异的座椅起到装饰作用
7. 客房

Hotel Riva

河岸酒店

Location: Hvar, Croatia
Completion: June 2008
Designer: Jestico+Whiles
Photographer: Ales Jungmann

地点：克罗地亚 赫瓦尔
翻新时间：2008年6月
设计：Jestico + Whiles 设计事务所
摄影：亚历克斯·荣格曼

BACKGROUND

The original hotel, then called the "Slavija", occupied two existing buildings, built in the typical Croatian Medieval style with a stunning stone façade, on the harbour front of Hvar town on the Adriatic island of Hvar in Croatia. The hotel's original construction, including fluted columns and stone walls, is frequently exposed.

OBJECTIVE

The design brief from the client was to create a chic, stylish and contemporary world class hotel on the island of Hvar, also known as the Lavender Island, to add to and compliment the newly refurbished Hotels Amfora and the future Hotel Adriana (also a Jestico + Whiles development).

SOLUTIONS

It involved total re-planning of the hotel, extensive structural work and the installation of a new heating, air-conditioning and ventilation system as well as new kitchen spaces. The hotel had to be brought "into" the 21st Century and this was a major change.

The attention grabbing wall treatment and other slightly risqué design elements, help to reposition the hotel as an alluring and desirable destination for the modern jet-set. However, the main façade was untouched due to planning conditions and was totally refurbished with the inclusion of new windows and doors etc.

The hotel includes bespoke wallpaper decorated with large-scale drawings of nudes specially created for the project by Jestico+Whiles. Once inside the bedrooms, the design is brave and uncomplicated. Blocks of red are combined with images of vintage film stars screen printed onto cotton fabric backdrops to the beds. A local stone was sourced for expanses of wall cladding combined with beige sandstone and slate flooring and a limited colour palette of red and black for the furnishings. A vibrant red wall faces the bed and is the only colour that has been used inside the rooms. A simple glass screen wall framed in timber is all that separates the bedroom from the En-suite making the room feel much larger than it really is. The bathroom sinks, also in the vibrant red faces the bedroom and the window offering a clear view of the water. The glass screens to the toilet have an etched glass pattern to provide a little more privacy. There is no room for modesty in these bedrooms. The luxury suite with its own private terrace is a spectacular space to unwind with a gin and tonic and watch the sun set on Hvar town.

In addition to the 45 rooms and nine suites, Jestico+Whiles has designed a new hotel bar with a backlit, green, etched glass counter and a teak veneered top. A fusion restaurant is located on the other side of reception with further seatings on the enlarged terrace.

1. Exterior and outdoor rest area before renovation
2. Lobby before renovation
3. Restaurant before renovation
4. Refurbished exterior

1. 酒店翻新前外观及室外休息区
2. 酒店翻新前大堂
3. 酒店翻新前餐厅
4. 粉饰后外观

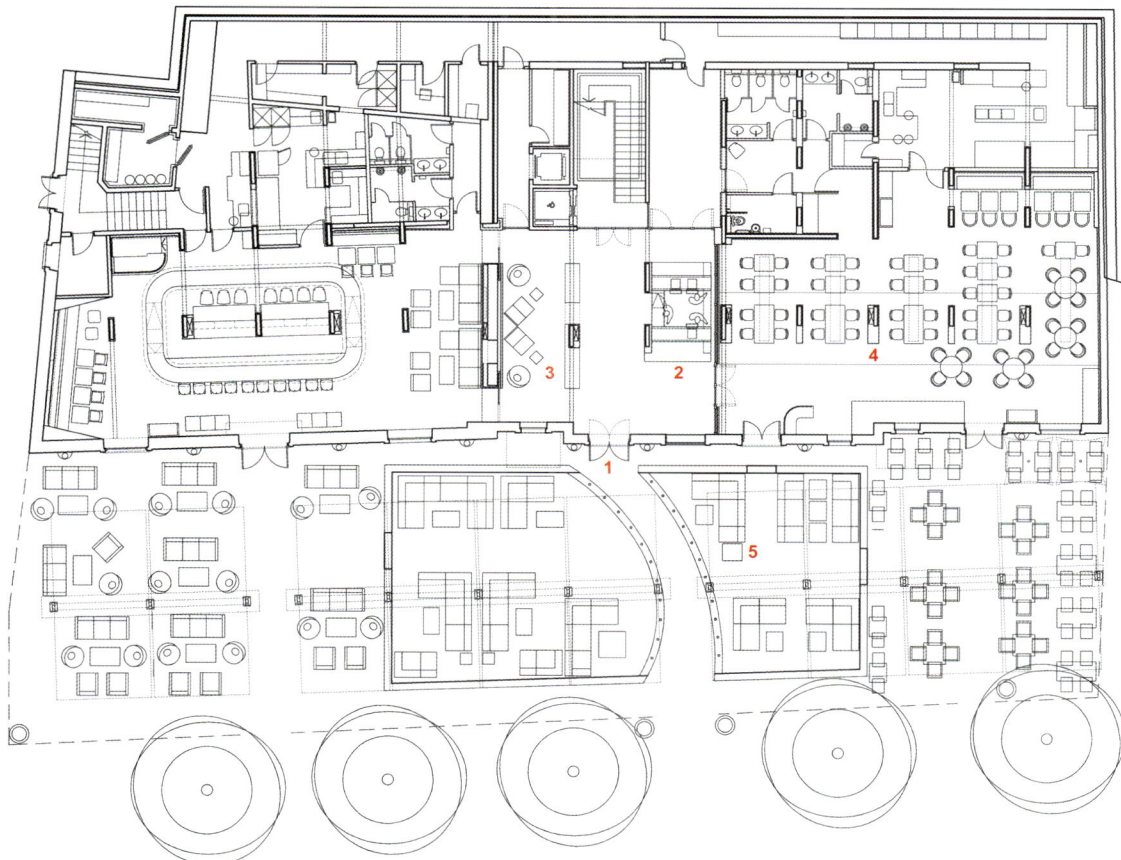

1. Entrance
2. Reception
3. Lobby lounge
4. Restaurant
5. Courtyard
1. 入口
2. 接待台
3. 大堂休息区
4. 餐厅
5. 庭院

5

背景

酒店前身为"塞尔维亚酒店",坐落在亚得里亚岛赫瓦尔市港口(克罗地亚的赫瓦尔岛通常被看作是亚德里亚海的蒙特卡洛——汽车品牌。这里受有钱人士的青睐,知名的、迷人的豪华游艇在赫瓦尔岛上随处可见)。两幢建筑内,美丽的石头外观突出了克罗地亚中世纪风格。原有建筑已有20多年未经任何修缮。

目标

客户的要求是在赫尔瓦岛上打造一个时尚、现代的酒店,以与附近新建成的Amfora酒店及即将建设的Adriana酒店相互呼应。然而,这一酒店又与其他酒店不同,需将不同的风格融为一体。因此,设计师目标为打造一个典雅、独特、活力十足并足够时髦的酒店,借以迎合那些开着豪华游艇入住酒店的客人的现代生活方式。

过程

这是一个大规模的工程,需要将酒店整体翻新,包括空间格局重新规划、结构框架改建、安装新的供暖、制冷及通风系统、打造全新的厨房空间等。酒店需满足21世纪风格,这对于设计师来说无疑是巨大的挑战。在酒店格局重新规划的同时,结构改建须严格按照已有的条件出发,这方面问题不大。主要问题是如何在现存房间的高度限制下安装现代化的采暖通风与空调系统;另一问题即为设计团队需与当地政府沟通,是否可以在酒店前面建设新的结构。

该建筑是中世纪时期的风格,酒店原有的框架,包括瓦楞柱和石墙常年裸露。从水边广阔的平台步入酒店,历史的建筑材料与这里时尚、鲜艳色彩的家具和装置并列,吸引年轻的或有颗年轻的心的客人。J+W事务所特意为该项目创作了大型的裸体画装饰室内墙壁。吸引人注意力的墙面处理和其他一些稍微有些暴露的设计元素将酒店重新定位为迷人的、令人向往的现代富人的度假村。

酒店内的明星画墙纸是由Jestico+Whiles事物所专门打造的。步入客房,设计大胆且不复杂:运用了红色块,经典的电影明星剧照印在棉布料上作为床头上部的背景墙。当地的石材被用在墙壁上,并同时采用了米色的砂石和石板做地面材料。室内的设施只用了黑色和红色两种色调。浴室的手盆也是红色的。卧室和浴室间的玻璃墙给人以空间扩大之感。没有门的间隔,只是在玻璃上有几条半透明的磨砂带,给人一些私密感。

除了45间客房和9间套房外,J+W事务所还为酒店的新酒吧设计了背光照明的绿色蚀刻玻璃立面和柚木台面的吧台。接待台的另一端是一个一体化的餐厅,它和一个扩大的平台相连,平台也是餐厅的室外就餐区。

5. Outdoor terrace by night and the charming sea view in front
6. New hotel bar with a backlit, green, etched glass counter and
   a teak veneered top
5. 夜色下的室外休息区以及不远处迷人的河流景致
6. 木质台面一直延展到吧台上方天花的一侧，内置灯饰将吧台正面映
   成绿色，闪闪的珠帘从吧台上方垂挂下来

7

7. Bespoke wallpaper decorated with large-scale drawings of nudes specially created for the hotle
8. The backdrop wall to the bed clad with a complete expanse of local stone
   and images of vintage film stars screen printed onto cotton fabric
9-10. Large blocks of red used to highlight the brave design
7. 走廊两侧墙壁上画着各种姿势的人物素描画，下部分用亮红色粉饰，打破了白墙
   的单调感
8. 床的背景墙采用当地整块石材饰面以及印在棉布料上的经典电影明星剧照装饰
9-10. 大片红色的运用突出了设计师大胆的理念

Hotel Murano

穆拉诺酒店

Location: Tacoma, Washington, USA
Completion: 2008
Designer: Corso Staicoff, Inc.
Photographer: John Clark, David Phelps
地点：美国 华盛顿州 塔科马
翻新时间：2008年
设计：科尔索·斯泰科夫设计公司
摄影：约翰·克拉克 、大卫·菲尔普斯

OBJECTIVE

Given the commission to renovate the weathered chain hotel, the design team looked to the flourishing local art community for influence. In the same manner, they wanted to link the hotel to the community using glass as its vehicle.

SOLUTIONS

Taking the hotel lobby back to its original, pure architecture offered a harmonious environment for art glass installations. It was critical that the backdrop be minimal and neutral to allow the art to be the focus.

Art glass is incorporated into the architecture - the front desk, entry doors, lobby chandelier and public restroom sinks were all created by internationally known artists. A cool blue glow floods the entry and lobby bar through stacked glass walls while the bar itself has a slump glass counter, illuminated from within.

Each of the 21 guest floors is dedicated to a single artist, featuring work displayed behind a customised etched glass wall engraved with artist quotes and commentary. Photographs and sketches along the corridor and in the guest room shed light on the artistic process.

Hotel Murano's rooms are individually unique while sharing an extensive attention to detail. Rooms feature hand-blown glass bedside lamps and glass-topped vanities alongside custom-designed furniture.

1. Lobby before renovation
2. Fireplace area before renovation
3. Corridors before renovation
4. Lobby with boat-shaped decorative structures suspended from the ceiling
1. 酒店翻新前大堂
2. 酒店翻新前壁炉区
3. 酒店翻新前走廊
4. 翻新后的大堂内部犹如小船一般的装饰物从天花上垂悬下来，趣味十足

5

目标

设计的目标即为将一座饱经风霜的连锁酒店翻新，设计小组从当地盛行的艺术社团那里获取灵感，同时用玻璃艺术做直通车，将酒店与周围的环境联系起来。

过程

原有的酒店大堂，纯净的建筑风格为安装玻璃艺术品提供了一个和谐的氛围。简约的背景以及中性色彩装饰，急需让艺术成为此处的焦点。

玻璃艺术品与建筑融为一体，接待台、入口门厅、大堂吊灯及公共卫生间水槽处的艺术作品全部出自国际知名艺术大师之手。淡雅的蓝色灯光穿透层叠玻璃墙，洒满了入口和大堂酒吧。吧台处也装饰有垂直的、彩色条纹玻璃，照明灯饰安装在吧台里面。

客房共21层，每层均由一位艺术家的作品来装饰。作品展示在预制的刻字玻璃墙内，墙上展示艺术家的留言和解说。走廊两侧以及客房内，都布置了照片和素描画装饰，突出了艺术气息。

酒店内的每一间客房都别具特色，而细节则构成了设计的共同特色。手工吹制的玻璃床头灯和玻璃面的洗手盆，以及定做的家具更增添了其独特的韵味。

5. The pure white sculpture in front of the lobby lounge just like a model
6. Splendid chandelier hung from the ceiling as well as the specially designed backdrop to the sofas being eye-catching features

5. 大堂休息区前面的纯白雕塑犹如模特，别具特色
6. 华丽的吊灯以及独特设计的沙发背景成为了休息区内两大特色

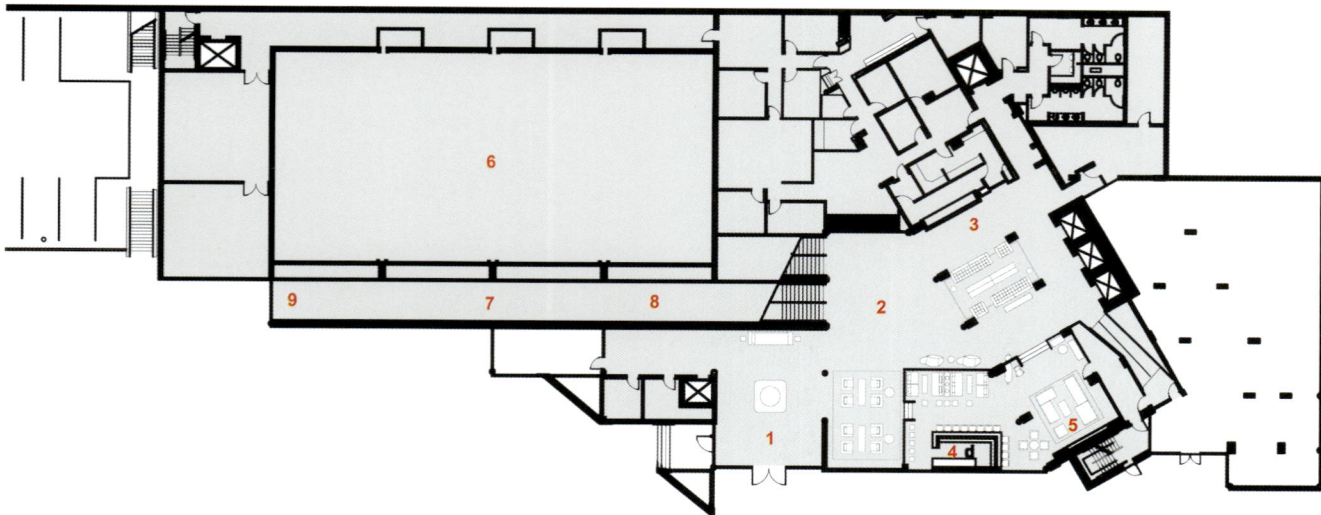

1. Main entrance
2. Lobby
3. Reception
4. Lobby bar
5. Fireplace sitting
6. Ballroom
7. Grand hall
8. Glass boat overhead
9. Entrance the first floor

1. 主入口
2. 大堂
3. 接待区
4. 大堂酒吧
5. 壁炉座位区
6. 舞厅（一层）
7. 大厅（一层）
8. 头顶上的玻璃船
9. 入口（一层）

8

7. Renewed fireplace area
8. Guestroom boasting simple style to create a homey atmosphere
7. 翻新后壁炉区
8. 客房设计简约，营造如家般的感觉

Hotel Palomar Los Angeles

洛杉矶帕罗玛酒店

Location: Los Angeles, California, USA
Completion: 2008
Designer: Cheryl Rowley Design, Inc.
Photographer: Cheryl Rowley Design, Inc.
Project Budget: $81 million
地点：美国 加利福尼亚州 洛杉矶
翻新时间：2008年
设计：谢丽尔· 罗利设计事务所
摄影：谢丽尔· 罗利设计事务所
预算：8100万美元

BACKGROUND

In the fall of 2005 a faded 19-storey building was tucked away on Wilshire Boulevard between Westwood and Beverly Hills. Aging and dilapidated, with low ceilings and a cramped, dark lobby, this Doubletree Hotel was an out-of-place eye-sore in desperate need of a complete renovation.

OBJECTIVE

Renovation and rebranding of a former Doubletree Hotel.

SOLUTIONS

Challenged to recreate and reposition this property the design team took its cues from its location: one of the world's premiere addresses amidst the iconic luxury high rises along the glamorous Wilshire Corridor. Developing an elegant design vocabulary comprised of a high-contrast graphic palette and an array of luxurious finishes, the design team has transformed the former Doubletree into a sophisticated and stylish Hotel Palomar with a distinctly residential feel, reminiscent of its neighbouring high-rise penthouses.

Guests enter on a wenge and white oak chevron-patterned wood floor that creates a dramatic exclamation point under chocolate custom carpets with carved cream-coloured tropical leaves. Backlit patterned glass panels lead guests to the macassar ebony and platinum leaf registration desk. In the lobby, warm platinum walls envelop the space, a soft palette to host the richness of macassar ebony and wenge millwork. The hues of taupe, chocolate and French grey come alive with accents of crimson, aubergine and hints of silver. Sumptuous textiles elegantly cover deco-inspired furniture shapes.

An overscaled decorative bronze mirror running the length of the lobby reflects the stunning ceiling-height rouge marble fireplace. The shimmer of suspended acrylic panels serves as a dramatic backdrop to inviting curved sectional seating. Touches of glass and polished chrome add sparkle and energy to the interiors and dramatic red lacquer walls, reminiscent of the lips of Hollywood sweethearts, draw guests into the elevator lobby and up to the guestrooms above.

1. Lobby restaurant before renovation
2. Corridor before renovation
3. Poolside area before renovation
4. Lounge area with fireplace
1. 酒店翻新前大堂餐厅
2. 酒店翻新前走廊
3. 酒店翻新前泳池休息区
4. 大堂休息区设计着壁炉

5

大堂内采用鸡翅木及白色橡木材质拼接地板铺设，为上面铺着的带有奶油色热带树叶图案的褐色地毯提供了完美的背景。背光照明玻璃板引领着客人一直走向带有铂金线装饰的黑檀木接待台。银色的墙壁将空间围合起来，为黑檀木及鸡翅木工艺品营造简约背景；灰褐色、褐色在灰色与淡紫色、紫红色与银色的映衬下，更显活泼；华丽的织品铺盖在艺术装饰风格家具上，别具一番特色。

一面大幅的铜镜贯穿整个大堂，将与天花等高的红色大理石壁炉映射其中。闪光的垂悬纤维塑料板构成了延展座区的背景，玻璃以及抛光铬材质带来了闪耀的动感气息，红漆墙壁让人不禁联想到女星魅惑双唇，吸引着客人走进来。

5. Reception and rest area in front of it
6. Bar area
7. Restaurant
8. Renewed corridor - red walls and stripe-pattern floor emitting fashionable feeling
9. Stripe-pattern floor seemingly as waves under the light
10.Suite

5. 接待台及墙面的休息座区
6. 酒吧区
7. 餐厅
8. 翻新后的走廊，大红色墙壁以及条纹图案地面散发出时尚气息
9. 条纹图案地面在灯光照射下好似摆动的波浪
10. 套房

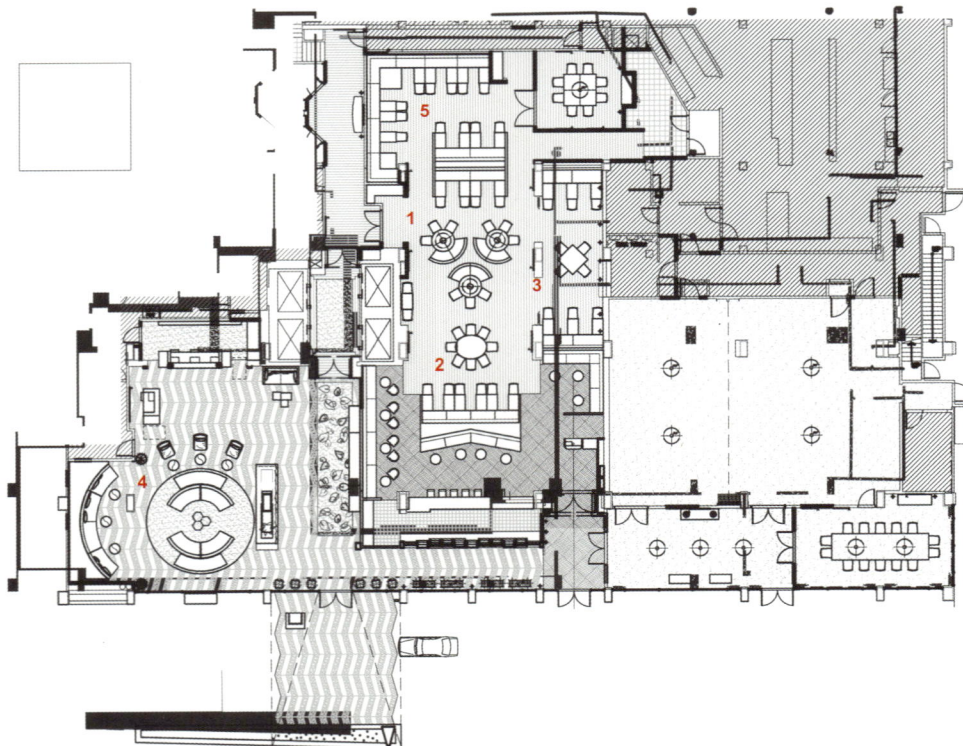

1. Lobby
2. Lobby bar
3. Reception
4. Outdoor pool
5. Restaurant

1. 大堂
2. 大堂酒吧
3. 接待台
4. 室外泳池区
5. 餐厅

11. Poolside area after renewing and the "lanterns" fixed on the wall being extremely distinctive
12. The fireplace in the outdoor area contrasting strongly with the surrounding blooming trees

11. 泳池休息区经翻新之后，固定在墙壁上的"灯笼"别具一番趣味
12. 室外休息区的壁炉与四周郁郁葱葱的树木形成鲜明的对比

Hotel EOS

EOS酒店

Location: Lecce, Italy
Completion: 2008
Designer: Luca Scacchetti Architects
Photographer: Marino Mannarini

地点：意大利 莱切
翻新时间：2008年
设计：Luca Scacchetti 设计公司
摄影：马里诺·曼纳里尼

BACKGROUND

The Hotel EOS in Lecce is a project of replacement building inside the old centre of Lecce. There existed a building, a two-storey house. On the same airbase was erected the new volume respecting the views of the adjoining buildings, but with greater heights to enhance the corner position. In this sense it is a recovery that emphasises an urban location that becomes a monument.

OBJECTIVE

The objective of the project is to represent the place in accordance with the contemporary design that has been entrusted to young architects from Lecce and young architects in the north Italy, through a partnership of the Politecnico of Milan.

Each has been appointed two rooms, creating a storey, a sum of differences and interpretations of the territory. The project is also double, and follows two different ways. Two pathways that have the same goal, however, the same objective.

SOLUTIONS

The first way concerns the architectural box, the wrapper. The façade in stone is carved roughly and randomly reproducing the contrast of the magnificent carved façades. It mimics the appearance, as under a sighted eye and without glasses, where the lines of the sculptures are not read to make space only for more light and shade.

The windows follow a complication of pseudo-baroque rhythmn (or very current), while a vertical cut, which corresponds to the scale, cracking the walls as the cuts in the extraction of the stone quarry.

Inside this contemporary representation of the Lecce Baroques, turns the second way of the project concerning the interiors. While the common areas reproduce symmetries, asymmetries and changes in tone, as in a seventeenth century music book, the rooms are as newsstand or chapels of the Basilica, and they are built as different part.

The former Continental Hotel has been completely renewed with new structures, new name, new façades covered with "leccese" stone. The 3-stars structure hosts 40 rooms, restaurant and lobby.

3

1-2. EOS Hotel building site
3. Façade in "leccese" stone of EOS Hotel
1–2. 酒店翻新前状态
3. 整修之后外观

4

背景

EOS酒店位于莱切老城中心一幢两层的建筑内，高度的增加突出了其所在的角落位置，使其成为这一区的标志性建筑。

目标

设计目标即为突出酒店所处的位置以及现代化风格。这一项目由来自莱切的年轻设计师及意大利北部设计师共同完成，每一组负责两个房间使自己的设计能够讲述独特故事的同时诠释出不同的地域特色。更为重要的是，虽然设计方式不同，但遵循同样的目标。

过程

首先是建筑外观的改造。表皮未经处理的石头随意镶嵌拼接而成，与那些精雕细琢的实质外观形成鲜明的对照，看上去给人一种模糊不清的感觉。上方特意雕琢的空格结构旨在引入更多的光线或是阴凉。窗户的设计遵循巴洛克风格，垂直的划痕犹如石材在采石场中便被划破一般。然后便是室内空间的设计，同样遵循巴洛克风格。公共区内对称与不对称样式共同存在，色彩不断地变化，客房则如同书报亭或是教堂一般，特色十足。

总之，原来的洲际酒店被彻底赋予新的面目——新结构、新名字、新外观。

4. Light shift between the hall, the wine bar and the living
5. "Espera" coffee and wine bar
6. Detail of the hall
7. "Sole, Mare, Jientu", room by Marta Picco
8. Bathroom of "Rure" room by Francesco Fiore and Lorenzo Spagnolo
9. "Cambra", room by Alessandro Giuri e Mario D'Aquino
4. 大厅、酒吧及休息区之间的光井
5. "Espera" 咖啡厅及酒吧
6. 大厅内细节设计
7. "Sole, Mare, Jientu" 客房
8. 浴室
9. "Cambra" 客房

6

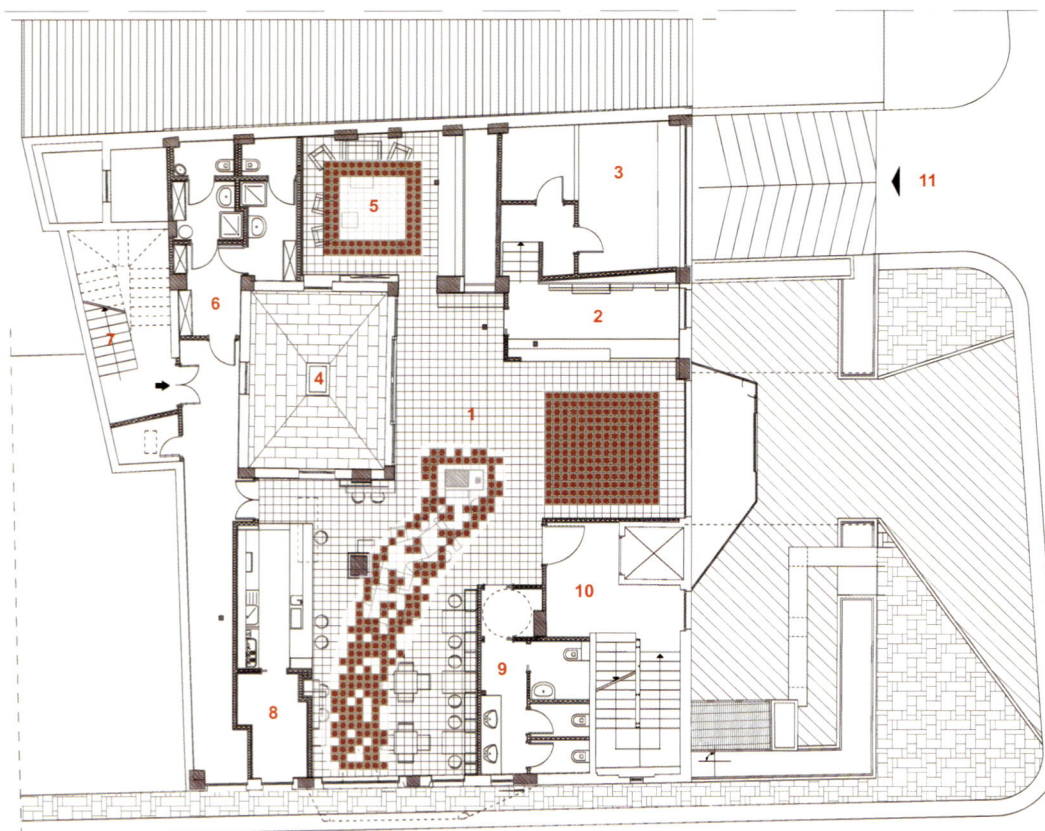

1. Hall
2. Reception
3. Office
4. Water garden
5. Living room
6. Service bathroom
7. Service staircase
8. Storage room
9. Bathroom
10. Stairs
11. Parking

1. 大厅
2. 接待区
3. 办公区
4. 水上花园
5. 起居室
6. 服务区浴室
7. 服务区楼梯
8. 储藏室
9. 浴室
10. 楼梯
11. 停车场

7

8

9

Newstay Vogue Hotel

张家港新世汇酒店

Location: Zhang Jiagang, China
Completion: Augest 2008
Designer: VJian Design Office: Song Weijian, Zhang Nan, Li Peng
Photographer: Song Weijian
Area: 6,200m²
地点：中国 张家港
翻新时间：2008年8月
设计：宋微建、张楠、李鹏/上海微建建筑空间设计有限公司
摄影：宋微建
面积：6200平方米

OBJECTIVE

"Newstay" is a new hotel brand. The designers were very familiar with the requirements of the client and guests for they begin their work at the inception of the project, including market position and VI design. Therefore, the result was very satisfying.

The original building, an old hostel built in early 19th century, was outdated in modern times. Designers proposed the concepts of "chic, natural and direct" to give the old structure innovative features.

SOLUTIONS

The original structure did not have a unified layout for being connected by four individual buildings constructed in different periods. In the small lobby, the space was casually separated and columns were arranged in chaos. To solve the problem, designers proposed a pond to unify the columns. Glass balls of different sizes were dangling above the pond, reflecting the light on the ceiling and leaves shadows on the pond to resemble the jumping notes. In the central area, a 6-metre-long communal table made of Padauk wood was placed for guests to relax and enjoy the natural atmosphere at the same time.

The guestrooms are not very spacious. The walls in the bathroom were removed to install mirrors, and the bathroom became larger and more transparent. Redundant decorations were demolished with only the necessary functions maintained for each structure. The beauty of each structure's function lies in the straightness and the order that it possesses and shows to people. The hardboard and concrete were kept as the original colour to show the charm of material. Soft furnishings, such as green fibre, grey curtains and vine structures create a warm and natural atmosphere.

There are two restaurants in the hotel, one Chinese restaurant on the ground floor and one Western restaurant on the third floor. In both two restaurants, the original elements were partially kept and then renovated with different styles. Thus, each has its own feature and at the same time combines with the whole hotel.

Just as the designers said: "Remodelling an old building is just like making friends with a cute but very stubborn lady. 'She' always challenges our nerves, but as the problems being solved one by one and finally a charming hotel is being realised, it is really satisfying. We can carefully enjoy and experience the beauty 'she' gives off."

1. General exterior view before renovation
2. Renewed frontage by night
3. Main entrance
1. 酒店翻新前外观
2. 夜色下翻新后外观
3. 主入口

2

3

4

目标

"新世汇"作为全新的酒店品牌，从设计初期就介入其中，包括市场定位以及酒店整体VI设计。设计师了解客人、业主的需求，也创造了他们共同想要追寻的感觉。

项目主体原为一处旧的招待所，建于20世纪90年代初，本身结构已经难以适应现代酒店的需要。根据现场情况，设计师提出"时尚、自然、直接"的设计关键词，赋予酒店全新的个性特征。

过程

因为是由四栋不同期建成的单独建筑连接而成，原建筑整体缺乏统一布局。大堂区域柱子凌乱，空间被无序的分割，显得狭小又混乱。为解决这个问题，设计师将密集的柱子通过水池连接为一体，在水池顶部悬挂下大小不一的玻璃球，玻璃球反射出顶部的灯光以及柱子和水池的光影，高低错落间如同跳动的音符；大堂中心位置放置了一张6米长的红花梨原木长桌，客人可以坐在这里喝上一杯咖啡，感受咖啡的香浓，感受淡淡木香和潺潺流水带来的自然气息。

客房面积不大，卫生间的墙面被拆除换上正面玻璃，空间变得通透起来；多余的装饰去掉，每个部件都简单到只保留它自身的功能，功能之美于是直接、有序的展现于人们面前；材质上保留了密度板的原色和水泥的原色，让通常作为配角的材料直接"说话"，告诉人们它们的组合原来可以有多完美、多和谐；软装部分，绿色绵麻、银灰色遮阳布以及各种藤质用品的搭配，营造出温馨、自然的氛围。

酒店一层设有中餐厅，四层设有西餐厅。两个餐厅都在各自空间内保留了一些原建筑的元素，再根据各自特征进行独立设计。既保持了各自风格又与酒店整体设计相互呼应。

"老建筑的改造，对设计者来说就像面对一个可人但又固执、爱找麻烦的女郎，她不断给我们'制造'新的问题，挑战我们的神经。然而，随着问题一个个的解决，一个独特又充满魅力的酒店呈现在我们眼前，在这个新生的世界里，我们慢慢打量、细细品味着经由时光雕刻和创意设计所带给她的美丽。"

4. Lobby
5. 6-metre-long communal table made of Padauk wood in the centre of the lobby
6. Chinese restaurant on the ground floor
7. Western restaurant on the third floor
8. Private dining room in the Chinese restaurant

4. 大堂
5. 大堂中心6米长红花梨原木长桌
6. 一层中餐厅
7. 四层西餐厅
8. 中餐厅内包房

1. Entrance   1. 入口
2. Lobby   2. 大堂
3. Waterscape   3. 水景
4. Reception   4. 大堂前台
5. Lift lobby   5. 电梯厅
6. Fire escape   6. 消防通道
7. Service   7. 服务台
8. Lounge   8. 散座区
9. Private dining   9. 包厢
10. Guestroom   10. 客房
11. Kitchen   11. 厨房
12. Office   12. 办公室
13. Staff canteen   13. 员工餐厅
14. Changing room   14. 更衣室
15. Auxiliary   15. 辅助房
16. Security room   16. 保安室

The Mercure Eastgate Hotel

默丘里·伊斯特盖特酒店

Location: Oxford, UK
Completion: 2008
Designer: Blacksheep
Photographer: Gareth Gardener
Area: 415m²
地点：英国 牛津
翻新时间：2008年
设计：Blacksheep 设计公司
摄影：加恩斯·加德纳
面积：415平方米

OBJECTIVE

The project - which encompassed a new 90-cover restaurant (with private dining area), bar, lobby, lounge and toilet areas - necessitated a complete redesign of the ground floor footplate to optimise key revenue-generating areas for the client and completely refresh the hotel's offer for guests. The aim is to achieve a 4-star hotel.

SOLUTIONS

The idea for the new ground floor plan was to rationalise the spaces so that the scheme flowed easily from one space to the next, with easy and effective change from day to night-time use. The brief was to enhance the customer journey and seek to increase the dwell time of hotel guests, as well as looking to attract more discerning local guests to use the restaurant and bar facilities more frequently.

The kitchen was to stay in its existing location and the new public spaces had to work around this. There was also the opportunity to upgrade the space in accordance with DDA regulations, including the installation of a new disabled lift. Blacksheep's key decision was to switch the spaces around so that the restaurant was located at the front of the hotel, linking in to the new bar, lounge and reception area.

The reception area of the hotel was reconfigured to create improved focal points and circulation, removing a through-door into the former restaurant space to create a more open, lobby feel. The existing reception desk was moved to the right of the entrance, a more pleasing symmetry was created by its position opposite a large existing sandstone fireplace. The bespoke-designed desk is in black-stained timber with a very hard-wearing gold paper with a hammered finish at the front.

Blacksheep then created a new lounge and lobby area from half of the existing restaurant space, with the other half used as a dedicated bar. The boundaries between these sections can move between day and night-time usage. The ease of the transition between the spaces is underlined through the use of timber flooring throughout, from the reception right through to the bar.

Two level changes in the space were dealt with via a DDA-conforming lift and ramp, whilst the toilets were also reconfigured with new entrances.

1. Restaurant before renovation
2-3. Lobby lounge before renovation
4. Refurbished exterior
1. 酒店翻新前餐厅
2-3. 酒店翻新前大堂休息区
4. 酒店翻新后外观

5

目标

这一设计包括重新打造位于酒店一层的空间，新建一个可容纳90人就餐的餐厅（带有私人就餐区），酒吧、大堂、休息区及卫生间等区域，更好的利用空间以及为客人提供更完善的服务。设计的目标即为达到四星级酒店标准。

过程

设计师重新规划一层的格局，使其合理化的同时强调空间流畅性及功能性，实现不同区域的有效转换。设计理念以"强调客人旅程愉悦性"为主，借以延长他们在酒店留宿的时间，同时吸引更多的当地居民来此就餐或小饮几杯。

厨房保留在原来的位置，周围设计了全新的公共活动空间。同时，设计中遵循无障碍设计条例（DDA regulations），包括残疾人专用电梯的设置，从而提升了空间品质。餐厅被设置在正前方，与酒吧、大堂休息室及接待台连通。

接待台加以重新设计，以突出其作为"中心点"的地位，并将通向原来餐厅位置的一扇门拆除，打造通透、开放的空间感。原来的接待台被移到入口的右侧，与一直矗立着的砂岩壁炉相对，打造出完美的对称性。此外，接待台是特殊定制的，黑色木质结构采用耐磨金箔纸装饰。

随后，设计师将原来餐厅的一半空间打造成休息区及大堂，另一半则用作酒吧，空间界限通过白天及黑夜的功能变换而转换。而整个空间内全部采用木质地板，从接待台到酒吧形成了连续统一的氛围。空间层数的改变则通过残疾人专用电梯及坡道设计解决了其带来的种种不便，卫生间也重新设计并打造了新的入口。

5. Chairs of different colours, patterns and materials emphasising diversity of the lounge
6. Specially shaped sofa in the lobby being rather playful
5. 不同颜色、样式及材质的座椅彰显出休息区空间的多样性
6. 造型独特的沙发座椅增添空间趣味性

6

1. Main entrance
2. Bar
3. Wine store
4. Fine dining
5. Main dining
6. Female WC
7. Male WC

1. 主入口
2. 酒吧
3. 酒库
4. 精品餐厅
5. 主餐厅
6. 女士卫生间
7. 男士卫生间

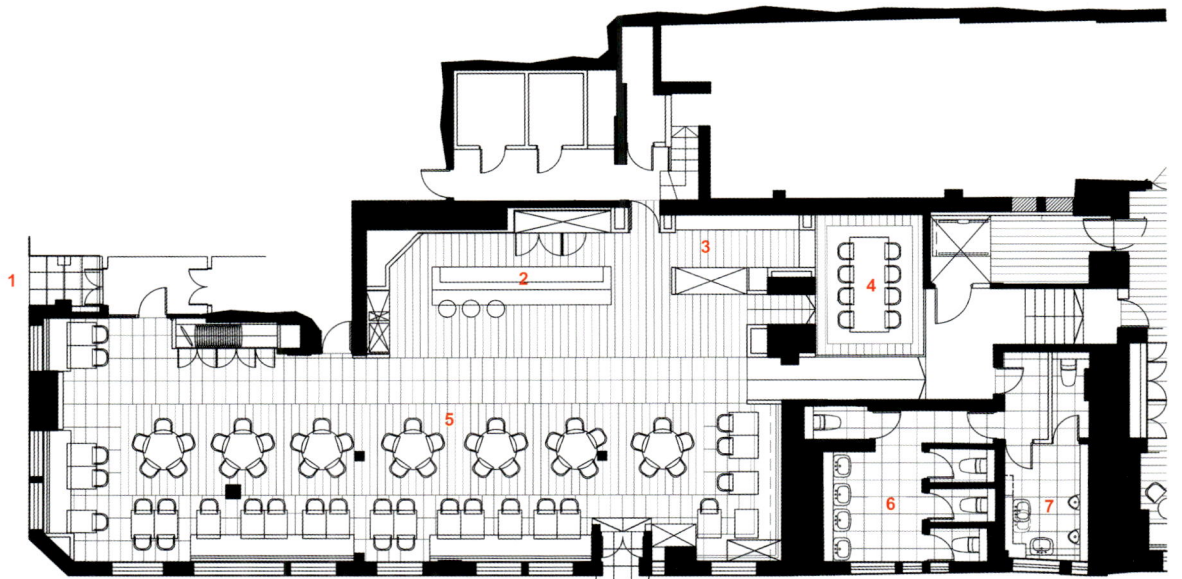

8

7. Corner of the lounge - the chairs and the floor corresponding each other in pattern and colour

8. Private dining room boasting simple yet elegant design

7. 休息区一角，座椅与地板在图案及色彩上相互呼应

8. 餐厅包房设计突出简约而雅致的风格

9. Bar area - light spreading on the counter being reflected to the ceiling, forming an enchanting visual effect
10. Restaurant stressing the theme of contrast between "black area"(black tables and black chairs) and "white area"(white tables and white chairs)

9. 酒吧区，洒落在吧台上的灯光反射到天花板上，形成了迷人的视觉效果
10. 餐厅设计强调对比主题，"黑色区域"与"白色区域"形成强烈的视觉对比

Yes! Fashion Hotel

悦色时尚酒店

Location: Tangshan, China
Completion: May 2007
Designer: LISPACE
Design Director: Jia Li
Photographer: Gao Han
Area: 5,200m²
地点：中国 唐山
翻新时间：2007年5月
设计：北京立和空间设计事务所
设计主持：贾立
摄影：高寒
面积：5200平方米

OBJECTIVE

The original building of this project is a local hotel with a bowling hall in the 1980s, and the character of the building had disappeared during use. The designer redefined the hotel as a small-size theme hotel, which has 48 guestrooms, one Chinese restaurant and one cafeteria. The investment is very limited, so how to control budget but still outstand design theme becomes the biggest challenge for the designer.

SOLUTIONS

The Chinese restaurant is reconstructed from the bowling hall. There are no windows in original space. The designer uses this disadvantage to transfer the restaurant as a "Chinese courtyard under the moon". Arc dome becomes the night sky. Custom-made modern Chinese lamps illuminate each table.

Renovation for guestroom building focuses on the blend of architecture and interior space. The designer keeps the main structure of the original façade and builds a simple glass box as main entrance. The courtyard that had been used as storage before has been changed to a breathing space. Cafeteria extends space into courtyard to bring scenery to each guest.

The area of guestrooms is limited, so how to impress all the guests becomes the first question for the designer. Standard guest rooms play with colours. Light goes through windows, and rich colours jump in night. There is a rectangular relative independent space in original building. The designer puts four theme guestrooms in it, and uses colours and patterns to describe four seasons. Simplicity and multi-function are characters of suite rooms, which gain the highest occupancy rate in this hotel. Rotating TV cabinet with mirror behind it divides the bedroom and living room. The wall of washroom is replaced by curtain and glass. Those elements give more free ways to guests to use their private space.

The renovation of Yes! Fashion Hotel considers details and saves energy. The hotel has become membership with several chain hotel groups after use. That is the value of design.

1. Exterior and main entrance before renovation
2. Lobby before renovation
3. Restaurant before renovation
4. Renewed main entrance by night
1. 酒店翻新前外观及主入口
2. 酒店翻新前大堂
3. 酒店翻新前餐厅
4. 主入口翻新后夜色下的景象

悦色时尚酒店

5

目标

本案建筑的前身是20世纪80年代的地方宾馆及保龄球馆，建筑自身的特点在使用过程中已消失殆尽。项目定位为小型主题酒店，包括48套客房，1个中餐厅，1个自助餐厅。资金预算十分有限，如何体现小而精的特点是设计师面临的最大挑战。

过程

中餐厅是利用原有的保龄球馆进行改造的。设计师利用封闭空间这一不利因素，反而将餐厅定义为"月色下的中式庭院"。弧形的穹顶成为繁星点点的夜空，定制的现代中式灯具照亮了每一个用餐的客人。"悦色时尚"便由此得名。

客房楼的改造重点在于建筑空间和室内空间的融合。对于建筑外观采用整合的手法，只保留简单统一的线条，入口中庭简洁的玻璃方盒子与老建筑自然的融为一体。内庭院在改造前几乎不为人所知，一直被当作仓库使用，设计师确将它作为整个酒店中最会呼吸的空间，自助餐厅的阳光和客房阳台的余晖使小小的庭院风景不断。

客房区域面积有限，如何使入住的客人印象深刻是设计师思考的问题。标准客房的设计体现了酒店的主题——悦色。通过客房中一盏盏的灯光，丰富的色彩在夜色中跳跃。客房楼中有一处相对封闭的井字空间，设计师将4套主题客房春、夏、秋、冬置于其中，通过色彩和图形的搭配诠释四季。春为花之色，夏为荷之美，秋为树之韵，冬为雪之景。套房的设计最为安静，也是入住率最高的房间。可旋转的电视机柜划分卧室与客厅空间，电视机柜背面为落地镜，兼顾了卫浴空间的使用。白色纱帘与玻璃隔断的结合，给予客人更自由的使用方式。

整个酒店的设计改造手法含蓄丰富，注重细节，节省能源。酒店在投入使用后，被多个连锁酒店组织纳为会员，这便是设计创造的价值。

5. The reception desk seemingly as a white box put in a green glass box, creating transparency and freshness
6. Oval pattern as the main theme in the washroom
7. Lobby bar
8. Attractive lighting fixture in the lobby bar
9. In restaurant, distinctive lighting fixtures splashing enchanting lights to create mesmerising dining atmosphere
5. 接待台犹如装在绿色玻璃盒子中的纯白盒子，营造十足的通透感及清新气息
6. 洗手间内椭圆造型构成了主要特色
7. 大堂酒吧
8. 大堂酒吧内的灯具格外吸引眼球
9. 餐厅内，各色灯光泼洒下来，营造了一个迷人的就餐氛围

7

8

1. Entrance
2. Reception
3. Restaurant
4. Private dining

1. 入口
2. 接待台
3. 餐厅
4. 包厢

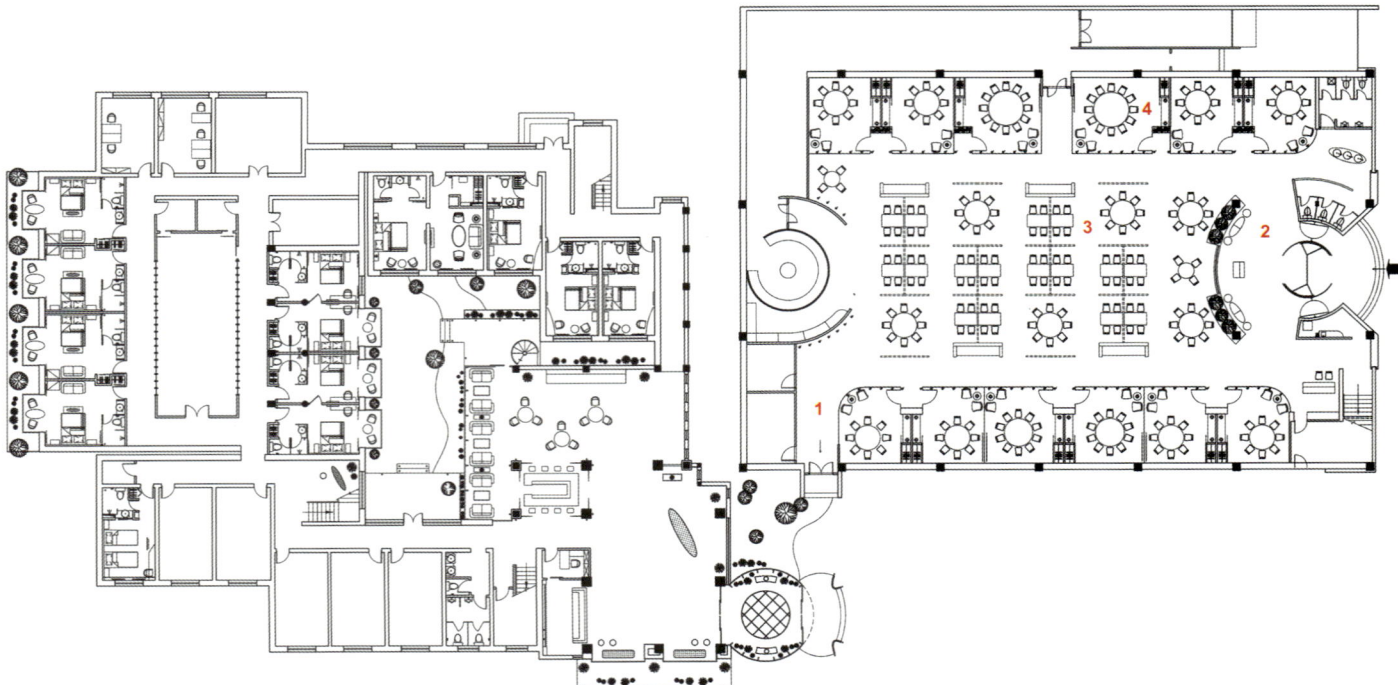

10. Guestroom designed with the theme of "spring"
11. Suite
10. 以"春天"为主题的客房
11. 套房

Mandarin Oriental Hotel

文华东方酒店

Location: Hong Kong, China
Completion: 2006
Designer: Lim.Teo+Wilkes Design Works Pte Ltd.
地点：中国 香港
翻新时间：2006年
设计：LTW装饰设计有限公司

BACKGROUND

The hotel set the style in September 1963. Its strategic location on the fragrant harbour lends itself as a central point for business and pleasure.

OBJECTIVE

"We find ourselves with an opportunity to mix 40 years of illustrious past into a glorious future. Refurbishing the Mandarin Oriental is a task involving refreshment and rejuvenation but without a change in philosophy. "

SOLUTIONS

For the guestrooms, designers propose two different design concepts in the desire to manipulate the opportunities provided by their locations within the building. Both schemes enlarge the space, one with ensuite glazing that allows views throughout the room, the other with sheer openness, while colour schemes have been developed to allow for either a warm and relaxing or cool and soothing atmosphere. A

vast collection of art and antiquities reside in the hotel and those will naturally be placed largely in the public areas. The Chinese influence in the public area artworks will enter the room by the use of craft. This gives the guest a first hand opportunity to encounter the intricacy of Chinese arts and crafts. The MO will never be big and flashy in regards to scale of public spaces, and despite the large number of rooms it has always maintained a sense of intimacy. A rich palette of Porto Oro marble stone mimics the previous black stone in the lobby, but increases the drama with its gold veining. Mandarin orange lacquer accents and plush velvet seating further extend the drama.

The captain's bar is just off the lobby, where the predominance of the red walls will also remain. Instead the designers have altered the navy accents, and brought in the sophisticated taupe colours into the carpet and leather upholsteries, along with some new updated tables so that rejuvenation has occurred to an old familiar face. The same goes upstairs in the Chinnery bar, where all is intact.

The new Cake and Coffee Shop is a 21st century interpretation of the colonial era… not literally, but in a contemporary sense. The designers have used a little rattan feel in the chair backs, although it has been cast in pewter coloured metal, to reflect back to the tin mines of South East Asia. The blackheart sassafras wall panels evoke tones of light and dark timbers simultaneously and give a warm envelope of café au lait colour. Details such as the louvred frosted glass divider screens are like the old bungalow houses of the steamy tropics. Bar stools look like a box of chocolates in the cake shop area.

Up on the 25th level, the former home to Vongs Restaurant will be reborn as Pierre, with stunning views of Hong Kong Harbour. Mirror

and glasses and reflective surfaces will make the room sparkle and allow Pierre to take centre stage.

Man Wah, the namesake Chinese Restaurant will also undergo rejuvenation. The designers are retaining the beautiful timber ceilings that were done in the years since its opening, and using Cambodian Indigo paper to cover the walls. For the Oriental Spa, they want to create an environment that reflects the Chinese roots.

In all the suites, they have tried to reinvent the suite with a stronger sense of self. The Corner Suites are a new addition to the hotel. The Macau Suite has a true inventors of the East meets West look. In the Tamar Suite, panelling and accents are to reflect the interiors of a private yacht. The Howarth Suite is traditional and English influenced and navy blue. The Meiji Suite is naturally Japanese, with a soft mustard ochre backdrop. The Mandarin Suite undergoes the most significant of changes and the floor is broken and the living room takes on a double level with soaring windows overlooking the harbour. There will be a greater emphasis on a large dining room, complete with bar. An entertainment room will be cocooning and comfortable with deep ochre and sand colours. Throughout the suite beautiful lacquer furniture with painting and inlay will be able to stand out as the backdrop becomes slightly more contemporary. The Master suite is increased in size, with a huge ensuite bath, the tub at window side, and an adjoining spa room for massage and exercise.

1. Cake shop before renovation
2. Guestroom before renovation
3. Suite living after renovation
1. 酒店翻新前面包房
2. 酒店翻新前客房
3. 翻新后套房起居室

4

背景

酒店于1963年9月建成，确定了自己的风格。位于美丽的港口边，酒店战略性的地理位置使其成为商业和娱乐中心。LTW设计公司因此也有了一个将40年辉煌的酒店历史和其壮丽的未来相融合的机会。

目标

整修文华东方酒店的任务是将其修复一新，重新注入活力，但不改变其原有的风格。

过程

设计针对客房提出两个不同的设计方案，以便巧妙地处理室内空间。两个方案都扩大了空间：一个是将与卧室相连的浴室加上玻璃隔断，使视线无阻隔；另一个是完全开放空间，用色彩主题的变换改变室内或温暖、休闲，或时尚、让人放松的氛围。这种装饰强调了位置感，有一种居家的氛围。

大量收集的艺术品和古董摆放于酒店之中，而且其中的大部分摆在公共空间里。客房延续了公共空间里的中式艺术，并用手工艺品来表现。在这，客人将首次接触、感受到中国艺术和手工艺品的精美。

文华东方酒店公共空间的尺度不大，也不浮华。虽然房间很多，但是一直给人一种亲密感。大堂内，色彩丰富的大理石替代了以往的黑色石，金色的纹理让这里增添了浪漫感。柑橘色的漆器和绒料的座椅强调并延伸了这种浪漫的色彩。

船长酒吧邻接大堂，红墙的主色调没有更改。我们改变了海军（深蓝）的主色调，地毯和皮沙发运用了有一定深度

的褐灰色，新换了一些桌子，是"旧貌换新颜"。楼上的Chinnery酒吧也和这里一样，所有的一切都互相呼应。

新建的蛋糕、咖啡店是对殖民地时期的现代诠释，但并不夸张。我们在椅背上加了一点藤条的感觉，但它是用锡金属铸成的，反映了东南亚产锡矿的事实。黑色的美洲檫木墙面板和浅色、深色的木板将咖啡店表皮装点如欧蕾咖啡似的温暖色调。细节设计包括有百叶隔断和磨砂玻璃隔屏，就像置身于热带地区雾蒙蒙的、古旧的平房里。在蛋糕售卖区，这里的椅子看上去就好像盒装的巧克力。

25层上，原来的Vongs餐厅被重新改造为Pierre餐厅，并总览香港港口的壮观景色。镜子和玻璃等反光表面使房间内熠熠生辉，让Pierre（大厨师的菜肴）成为焦点。

文华中餐馆也历经了改造。酒店开业时修建的精美的木顶棚被保留下来，在墙面上贴上了柬埔寨蓝墙纸。在东方温泉浴区，创造了一个反映中国根源的（文化）环境。

在所有的酒店套房中，设计师试图重新创造个性化的房间——角落套房是酒店新加建的；澳门套房中有中西合璧之美；他玛（基督教《圣经》故事人物）套房里，装饰的嵌板和格调好似私家游艇；Howarth套房是传统的英式风格，海军蓝是主色调；明治套房自然是日式风格，有芥末黄和赭色的背景（图案）。文华套房历经了最大的改造，地面打通，和起居室形成越层，大面积的窗户俯望港口。设计的重点还放在大餐厅上，并配有酒吧。一间娱乐房如蚕茧一样舒服，是深赭色和沙子的颜色。整个房间中，精美的上漆家具，还有

绘画、嵌入的彩色图片是突出的背景展示，更具有现代的风格。主人套房的面积扩大了，并带有一个大的浴缸，放在窗前，一个温泉室与其相连，在那里客人可以享受按摩服务，还可以锻炼身体。

4. Lobby lounge
5. Cake shop with a new look
6. Café
7. M Bar
8. Pieere Restaurant
9. Pieere Restaurant private dining
10. Man Wah Chinese Restaurant
11. Spa reception
12. Suite living
13. Suite bedroom
4. 大堂休息区
5. 翻新后的面包房
6. 餐吧
7. "M"酒吧
8. "Pieere"餐厅
9. "Pieere"餐厅私人就餐区
10. 中餐厅
11. 水疗接待区
12. 套房起居室
13. 套房卧室

9

1. Reception
2. Lounge
3. Bar
4. Cashier
5. Bell captain
6. Concierge

1. 接待台
2. 休息区
3. 酒吧
4. 收银台
5. 领班室
6. 门房

JW Marriott Chicago

JW万豪酒店

Location: Chicago, Illinois, USA
Completion: 2011
Designer: DiLeonardo International
Photographer: Barbara Kraft, Grant Kessler
Area: 51,097m²

地点：美国 伊利诺伊州 芝加哥
翻新时间：2011年
设计：迪利安达国际
摄影：巴巴拉·卡拉夫特 、格兰特·凯斯勒
面积：51097平方米

OBJECTIVE

208 South LaSalle is a historic landmark located in the heart of Chicago's Financial District. A Design Challenge was to take the hotel's stately architectural styles of Daniel Burnham and blend this with a sophisticated elegance of today's modern classic style.

SOLUTIONS

The inviting warmth of the Lobby is a fusion of grand scale; Burnham-inspired details, rich colour and a stimulating atmosphere. The lounge is a luxurious living room tucked behind the elegant stairway in the Grand Hall of the Lobby. The warm inviting glow of the fire pulls one in and is enhanced by burgundy marble cladding the adjacent bar. Communal tables in Wenge wood and opulent white marble cladding create the envelope of the space. The Marriott Great Room layout is a flexible setting for business or social events. The multi-functional settings can be used for individual business work, community connectivity working at the communal table or socially relaxing in the lounge area.

As the guest moves to the second level, whether by the grand stairways or by the escalators, they are drawn to the magnificent lighting fixtures that hang majestically throughout the lobby. The marble-clad columns soar 30 feet high culminating in a dramatic ceiling design.

Approaching the grand ballroom the guest is met with groin vaults and a rich layering of architectural details. Grand chandeliers reflect luxury and warmth. The prefunction area at the end of the ballroom backs up to a wine wall subtly offering glimpses of the Italian Florentine Steakhouse.

The Restaurant is a fresh design that blends urban with Italian country style. The handcrafted walnut bar is highlighted by Italian ceramic flooring in the bar area. Floor-to-ceiling shelving displays restaurant utensils, cook books, and dry goods continuing the open kitchen experience throughout the perimeter seating at the bar as well as in the main dining. Slivers of lighting wash down over the guests creating a subtle mood in the main dining that is set off by the brilliant intensity of the artistic graphics placed around the perimeter.

The rich tones that pulled you into the Lobby area create the colour palette for the guestrooms. The lavish gold and brown tones will keep one warm on those cool Chicago nights and the luxurious red accent will make one feel glamorous when getting ready for a night out. The handcrafted furniture of the Guestrooms continues to bring in the subtle architectural details that were fluent throughout the public spaces, accented by custom decorative hardware. These pieces create an upscale and timeless experience in the historic Burnham tradition.

This is not just another experience in another hotel; this is a night on the red carpet with the luxuries and comforts of home.

1. Elevation of the original office building
2. Existing lobby
3. Existing grand hall
4. Refurbished elevation and hotel sign
1. 原办公大楼立面
2. 原有门厅
3. 原有大厅
4. 改建后立面及酒店标识

5

目标

酒店位于芝加哥金融中心区内，其建筑曾是历史地标。设计面临的任务即为保留其庄严的丹尼尔·伯纳姆建筑风格，同时与现代样式的高贵典雅相结合。

过程

大堂内凸显高贵，同时又不失温馨祥和，伯纳姆风格细节、丰富的色彩与活力十足的氛围相互融合；休息区内，豪华的起居室"退到"楼梯的后方，温暖的火光吸引着客人进入，旁边大理石饰面的吧台格外引人注目。这里是酒店内规模最大的空间，格局灵活，可用于商务会议或举办各种活动。客人无论是走楼梯或乘电梯上到二层，都会情不自禁地被大堂内华丽的灯饰而吸引。大理石梁柱高达10米，一直耸入到天花处，形成了独特的屋顶装饰。

走近宴会厅便会发现拱形的天花以及丰富层叠的细节装饰，富丽堂皇的吊灯彰显华美与温馨。宴会接待区位于后方与葡萄酒墙壁相邻，可以瞥见意大利牛排屋的景象。餐厅内风格清新，将美国都市特色与意大利乡村气息结合，手工制作的

胡桃木吧台在意式瓷砖地面的映衬下更加突出。落地橱柜内展示着餐厅的各种器具、烹饪书籍以及干货。酒吧四周的座区或者是主餐厅内，客人可以尽情享受开放式厨房体验。银色的灯光拂过脸庞，让人心旷神怡。

大堂内的绚丽色彩同样体现在客房中——大量的金色以及褐色让客人在芝加哥寒冷的夜晚中感受温暖，而艳丽的红色更让那些在夜晚外出的客人感觉心情愉悦。手工制作的家具将公共空间内的细节装饰全部照搬过来，营造出一种高雅永恒的体验。

JW万豪酒店为客人打造独一无二的体验，在感受红毯之夜的雍容华贵的同时，更能体验到家的温馨舒适。

5. Lobby bar - red carpet and sofas echoing each other
6. Grand stair leading upstairs at the lobby reception highlighting the elegance and sumptuousness of the hotel
7. Grand foyer of the lobby
8. The grand ballroom with groin vaults and a rich layering of architectural details as well as grand chandeliers reflecting luxury and warmth

5. 红色地毯与沙发相互呼应
6. 接待区通往上层的楼梯，突显出酒店的豪华与雅致
7. 大厅
8. 宴会厅设计有拱形的天花以及层叠的装饰细节，华丽的吊灯彰显华美与温馨

8

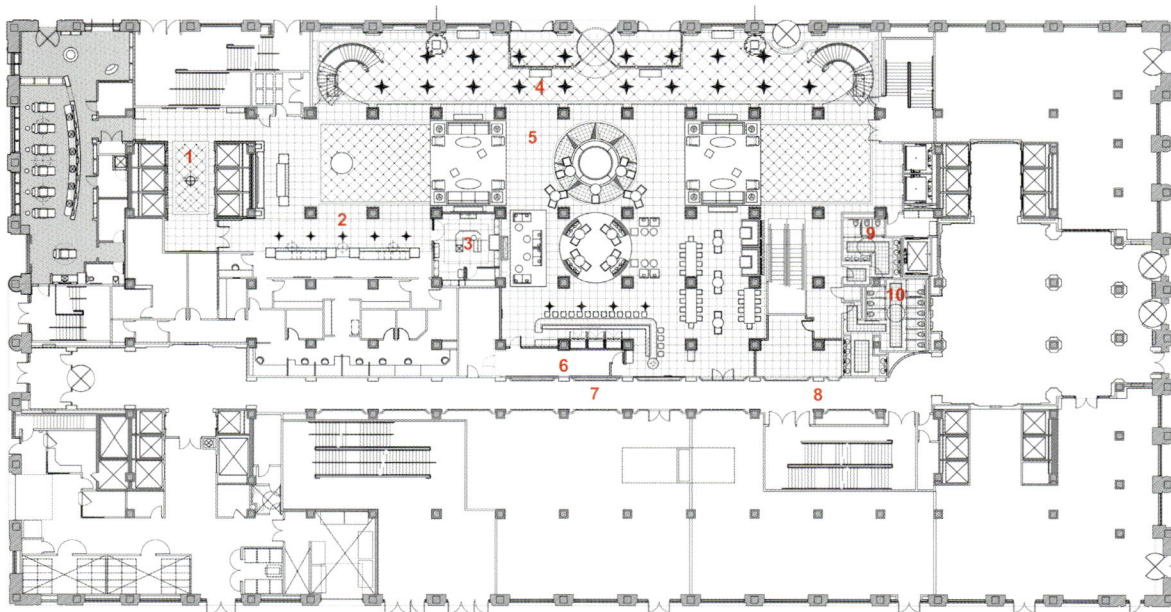

1. Elevator/lobby
2. Reception
3. Gift shop
4. Gallery
5. Lobby
6. Storage/pantry
7. Retail corridor
8. Men's restroom
9. Storage
10. Women's restroom

1. 电梯间
2. 接待台
3. 礼品店
4. 画廊
5. 大堂
6. 餐具储藏室
7. 零售店
8. 男士卫生间
9. 储物区
10. 女士卫生间

9. Executive Lounge as a luxurious living room creating a homey atmosphere
10. Florentine Dining Room
11. Semi-private dining
9. 休息室犹如豪华客厅，营造居家氛围
10. 餐厅
11. 半开放餐厅

10

11

12. Suite office
13. Typical guestroom
14. Suite bedroom
12. 套房办公区
13. 标准客房
14. 套房卧室

13

14

The Boundary Hotel and Restaurant

邦德里酒店

Location: London, UK
Completion: 2010
Designer: Sir Terence Conran and Conran & Partners
Owner: Prescott & Conran
Photographer: Paul Raeside
地点：英国 伦敦
翻新时间：2010年
设计：特伦斯·康兰&康兰设计公司
业主：普雷斯科特&康兰设计公司
摄影：保罗·雷塞德

BACKGROUND

The Boundary Hotel and Restaurant project involved the refurbishment and extension of a listed Victorian industrial building in Shoreditch, East London. Left unoccupied for many years, parts of the building's structure had become slightly dilapidated; the original 1893 configuration, however, had barely changed since its original inception.

SOLUTIONS

Boundary Rooms:

Over the course of a 2-year period, the layout of the original building has been altered by Conran & Partners to incorporate 2 new floors and a rooftop garden, clad in pre-patinated green copper and timber brise soleil. Care has been taken to retain the most charming of the building's original features; the distinctive brickwork, the large sash windows and industrial design light-wells. On the first and second floors there are 12 spacious guest bedrooms. On the third floor and fourth floors there are 4 duplex suites.

The boutique hotel comprises 17 bedrooms, 6 of which are duplex suites with views of the city. Each of the bedrooms in the hotel has an individual design, with some rooms expressing particular design movements or influences and others created by contemporary design figures including Sir David Tang, Priscilla Carluccio, Vicki Conran, Polly Dickens and Terence Conran. Bathrooms are also individually styled; some to suit the specific designers, others are wet rooms; classic Czech & Speake fittings are used in a few and more modern variations including the Japanese TOTO combined bidet lavatory.

Boundary Restaurant:

A formal restaurant in the lofty, daylight filled basement serving high quality, contemporary food to hotel guests and the local community in impressive surroundings. Dining tables laid with fine white linen, silver cutlery and table service from Terence Conran's Monno range are surrounded by chairs that have been sumptuously upholstered in deep red and indigo velvet. All furniture for the restaurant was designed by Terence and made by the sister-company Benchmark. In classic Conran style, a glass window along one side of the large central restaurant reveals the kitchen and crustacea bar, allowing diners to watch the skilled team at work preparing their food. The restaurant and bar floor is laid with French Bleu de Savoie marble, arranged in a bold geometric pattern.

Carefully lit displays of "objets trouvés" adorn the original exposed brick-work walls, alongside specially commissioned artworks and exhibits, including a large colourful three-dimensional piece called "Breaking the Boundary" by Richard Smith. There is also a decorative lighting feature which hangs from the ceiling, running the length of the restaurant. The piece is reminiscent of a flying carpet and serves to break the vast height, bringing intimacy and warmth to the space.

In addition, the ground floor is occupied by the Albion café and food retail area, serving robust, simple British food as well as take away food and drinks.

1. Exterior view of the original industrial building
2. Exterior view after being refurbished
1. 改造前原有工业建筑外观
2. 改造后酒店外观

背景

邦德里酒店位于东伦敦Shoreditch大街一幢维多利亚时期的工业大厦改建及扩建而成。大厦已废弃多年，部分结构因年久失修而备显破败。尽管如此，这一建于1893年的大厦在造型上却少有改变。

过程

历经两年的时间，设计师将建筑原有的格局进行了改变，双重斜屋顶彻底拆除，并加建两层，屋顶新建了花园，外观采用绿铜及木条镶嵌，内部设计着双人套房以及双层高度的起居室。其中原有建筑的部分特色被保留下来，如独特的砖石结构、上下推拉的窗户以及工业建筑风格的光井等。

酒店内共有17间客房，其中二层及三层共12间，四层及五层共5间，所有房间中共有6间为双人套房，并可观赏城市风光。值得提到的是，酒店内的每一间客房都具有自己的特色，有些在风格上突显某些设计运动或其产生的影响，有些则洋溢着现代特色。浴室的设计也各具特色，其中一些由专业设计师打造，另外一些为全封闭式的。

餐厅

餐厅位于地下室内，为酒店客人及当地居民提供高品质的美食。餐桌上铺设着精美的白色亚麻布，摆放着银制餐具，四周是陈列着深红色及靛蓝色丝绒靠垫的奢华座椅。所有的家具全部是专门设计的，餐厅一侧的大玻璃窗将厨房以及酒吧展露的一览无遗，客人在就餐的时候可以欣赏大厨的高超技艺。餐厅及酒吧的地面全部采用法国大理石铺设，并因其大胆的几何图案造型而格外引人注目。

古典风格的物品装饰在裸露的砖石墙壁上，在灯光的照耀下更显精致，加上专门制作的艺术品及展品别具一番特色。一个灯具饰品横跨整个餐厅，从天花上垂悬下来，让人不禁联想到"飞毯"，打破了空间的空旷感，进而营造出亲切而温馨的氛围。

此外，酒店一层为咖啡厅及食品零售区，出售简单的英式食品及快餐饮料等。

3. Rooftop rest area with fire place being the main feature
4. Carefully lit displays of "objets trouvés" adorning the original exposed brick-work walls
5. Restaurant in the basement with dining tables laid with fine white linen, silver cutlery and table service from Terence Conran's Monno range surrounded by chairs that have been sumptuously upholstered in deep red and indigo velvet
6. Reception area with vases of diverse colours and shapes making the whole space more playful
7. The Duplex Suite designed by Sir Terence Conran

3. 屋顶休息区，壁炉设计作为主要特色
4. 餐厅一侧的砖石墙壁上装饰着古典风格的物品，别具特色
5. 餐厅位于地下室内，餐桌上铺设着精美的白色亚麻布，摆放着银制餐具，四周是陈列着深红色及靛蓝色丝绒靠垫的奢华座椅
6. 接待区内各色的花瓶为空间增添了一丝趣味
7. 客房一角，插在墙上的树枝装饰格外吸引眼球

4

5

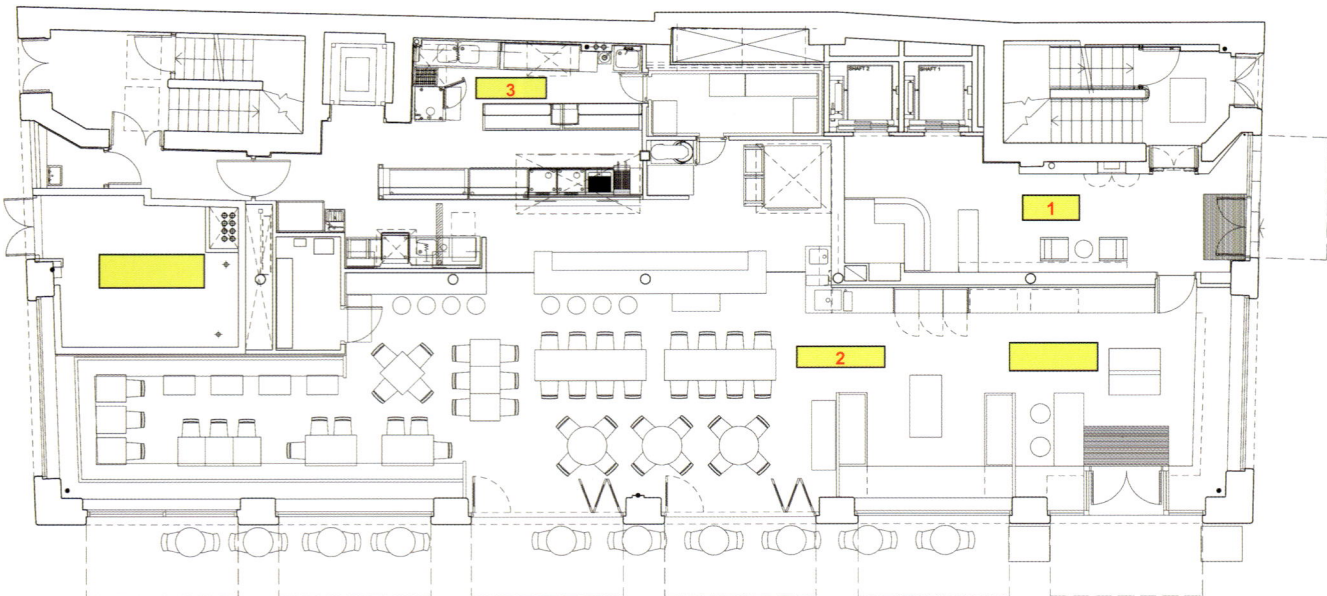

1. Lobby
2. Restaurant
3. Kitchen
1.大堂
2.餐厅
3.厨房

8. Guestroom with simple style is the main design theme, designed by Mies van der Rohe
9. Suite with plants patterns "climbing" from walls to ceiling and giving off lively and fresh feeling, designed by David Tang
10. Large block of glass mirrors enlarging the space visually, designed by Eileen Gray

8. 简约风格的客房
9. 套房内墙壁及天花上"爬满"了绿色植物图案，散发出生机勃勃的动感与清新气息
10. 玻璃镜子在视觉上使得空间更加开敞

Hotel Cosmo

丽悦酒店

Location: Berlin, Germany
Completion: January 2010
Designer: SEHW Architects
Photographer: Kathi Weber, Andreas Süß
Area: 5,600m²
Award name: AIT Award Extraordinary Interior Design
Award date: 2010
地点：德国 柏林
翻新时间：2010年1月
设计：SEHW 建筑事务所
摄影：凯西·韦伯、安德烈亚斯·瑞斯
面积：5600平方米
获奖名称：AIT奖 "杰出室内设计"
获奖时间：2010年

BACKGROUND

An office building, built only a few years ago and abandoned shortly after completion due to lack of tenants, became the shell into which the design hotel was poured.

SOLUTIONS

The exterior is a reference to the critical reconstruction of Berlin in natural stone, practically hiding the entrance, creating an introverted feel. Inside, SEHW Architects opened the ground floor into a bright lobby which is zoned by a dominant golden core consisting of service spaces. The lobby is cloaked in a light flowing curtain, blurring the threshold between inside and outside. The plan for the upper floors appropriates the existing office building grid and then emancipates itself in its design. The existing service core at the interior of the building is partially broken open to accommodate special functions such as a cigar lounge, spa, sauna, etc. The circular corridor is an orchestration of arrival and deceleration. Walls, ceilings and floors are coloured dark brown. Orientation is provided by specially designed pictograms, which are back-lit and lend atmosphere to the hallways.

One experiences a stark contrast upon entering the bright and generous rooms from the dark corridor, an explosion of wide views across the city or into the inner workings of the Ministry of the Exterior across the street. Inside the rooms there is no blast of colour or misunderstood youthfulness. Instead there are surprising details to be found everywhere; an interesting haptic experience is achieved with a catalogue of soft and rough, shiny and matte materials. Engaging visual axes are created through intelligent zoning and a bench that becomes a sideboard that becomes a wardrobe, developing along the wall. A sophisticated atmospheric lighting concept together with subtle design quotations of the opulent grand hotels of the past show that the Cosmo is the Berlin of today, without being retro. Cheers!

1. Ground floor of the original office building
2. First floor of the original office building
3. Second floor of the original office building
4. Exterior view of the hotel
5. Restaurant
6. Lobby
7. Lounge
8-9. Deluxe room
10-11. Standard room
1. 原办公建筑一层
2. 原办公建筑二层
3. 原办公建筑三层
4. 酒店外观
5. 餐厅
6. 大堂
7. 休息区
8–9. 双人套房
10–11. 标准客房

HOTEL

COSMO

5

背景

原办公建筑建于前几年，竣工之后由于缺乏业主而废弃，丽悦酒店便选址在这里。

过程

外观采用天然石材打造，突出了柏林地区重建建筑的特色。入口部分被"隐藏"起来，营造出含蓄内敛的氛围。室内改造上，设计师首先将一层全部打通，从而打造了一个明亮通透的大堂，包括服务空间。大堂内窗户采用窗帘"遮盖"，以此来模糊室内外空间的界限。上层空间的原有格局被更加合理化，原有的服务区被分割开来，用作吸烟室、水疗馆及桑拿房等。墙壁、天花及地面采用深褐色装饰，指向标识采用特别设计符号加以诠释，并通过背光照明，一直通往门厅。

从黑暗的走廊走近宽敞明亮的客房，给人一种强烈的反差。客房内可欣赏到城市的景色，设计上摒弃大片的色彩装饰，相反很多令人惊讶的细节随处可见。材料选用上同样突出对比性，柔和与粗糙、闪亮与暗哑共同存在，营造出趣味十足氛围，同时带有很强的触觉感。此外，通过对空间的合理划分，创造出完美的视觉轴线。沿着墙壁摆放的座椅"变化"成餐柜，之后又"形成"衣橱，特色十足。先进的气氛照明理念与精致的历史特色共同打造了悦丽酒店，同时展现出柏林的城市气息。

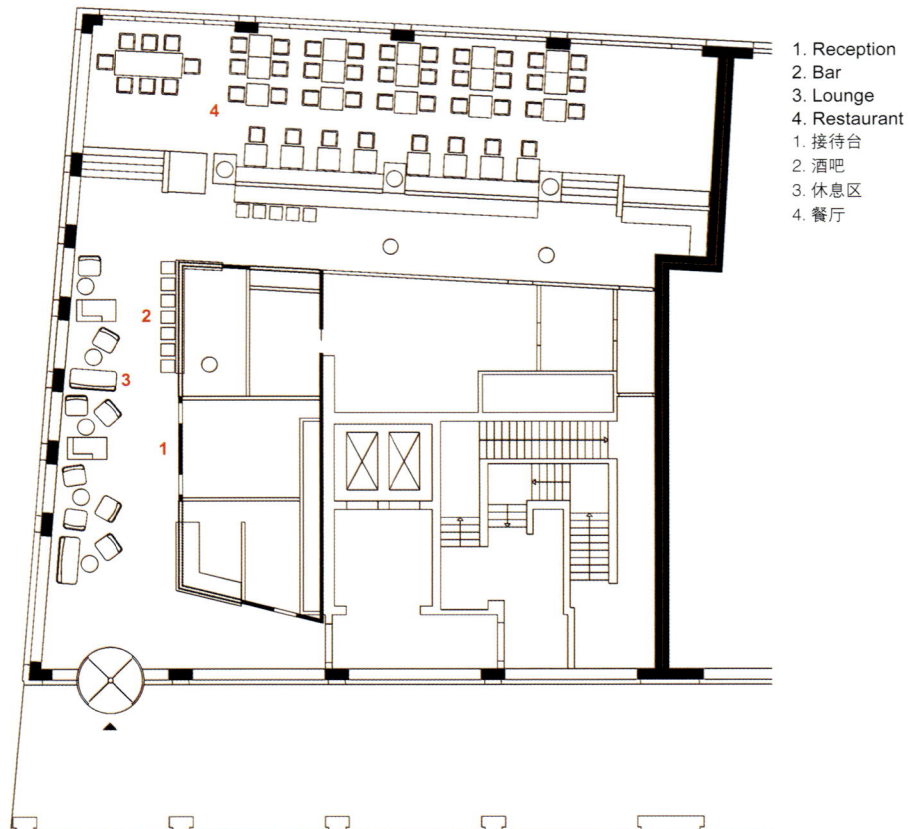

1. Reception
2. Bar
3. Lounge
4. Restaurant

1. 接待台
2. 酒吧
3. 休息区
4. 餐厅

Hotel Am Schottenfeld

绍特菲尔德酒店

Location: Vienna, Austria
Completion: 2009
Designer: Zeytinoglu Architects
Photographer: Falkensteiner Hotelmanagement GmbH, Brixen, Italy
Area: 3,250m²
地点：奥地利 维也纳
翻新时间：2009年
设计：Zeytinoglu 建筑设计事务所
摄影：Falkensteiner 酒店管理有限公司
面积：3250平方米

BACKGROUND

It was a new utilisation of old factory halls. The "Hotel Am Schottenfeld" was already 10 years old when it was refurbished in 2009.

SOLUTIONS

The lobby and restaurant areas were newly defined and a classy, cosy and spacious bar was established in the lobby. Dark leather seats contrast with silver curtains and – in combination with the warm wooden panels and the green-golden glass tables – create the elegant and at the same time cosy atmosphere which is expected from a hotel bar.

Two sunlit and friendly breakfast areas are separated from the lobby and the circulation areas by semi-transparent walls of wooden sticks, so that they can also be used for smaller parties or business dinners.

Between the hotel and the green facade of the loft wing lies a friendly green inner courtyard, below it is the location of a small conference centre with 4 new meeting rooms.

The new loft rooms are situated in the reconstructed and newly adapted area of the former printing plant Königsberg at Zieglergasse No.63. Room heights of up to 3.5 metres. Broad reveals and elegant dark oak wood floors give the place a special atmosphere. The rooms are divided into entry, working, and sleeping zones by floor-to-ceiling curtains in warm orange. The only important piece of furniture is the combination of bed and desk in the centre of the room. A lot of thought was put into the functional details of the puristic design of the loft rooms.

Together with the cosy new lobby area they create an elegant place right in the city centre for restless city travellers in need of recreation.

1. The original factory halls
2. Lobby before redesign
3. Welcome zone before redesign
4. Chairs of different colours (green and black) bringing touches of interest
1. 原有工厂建筑会议室
2. 原有工厂建筑大堂
3. 原有工厂建筑接待区
4. 不同颜色的椅子使空间充满趣味性

6

背景

绍特菲尔德酒店由一幢老工厂改造而成，在2009年翻新之前已存在十年时间了。

过程

大堂及餐厅区域是新规划出来的。酒吧设置在大堂内，空间开敞、风格时尚，黑色皮质座椅与银色窗帘形成强烈对比，暖色木板以及黄绿色玻璃桌营造了酒吧特有的典雅而舒适的氛围。

早餐区大堂以及过道区通过半透明木条墙壁隔离开来，光线充足并散发出友好的气息，这里更可用作小型会议室或者商务聚餐。

酒店及阁楼之间是一个栽种着绿色植物的小庭院，下方设计着小型会议中心，包括四个会议室。阁楼客房由原来的印刷厂改造而成，屋顶高度达3.5米。黑色的橡木地板营造出独特的氛围，房间由入口、工作室及卧室组成，不同的区域之间通过橙色落地窗帘隔离开来。室内最为抢眼的摆设即为卧床与书桌连成一体的家具。此外，阁楼房间设计在遵循简约理念的同时，更注重细节。

这一酒店为那些疲惫的城市旅行者提供了一个舒适、温暖的休憩之所。

5. Wooden table top and floor corresponding with each other and the orange table cloth brightening the whole restaurant
6. Rooftop guestroom
7. Loft guestroom

5. 餐厅内木质桌面与地板相互呼应，橘色的桌布则使得空间更加明亮
6. 屋顶客房
7. 阁楼客房

7

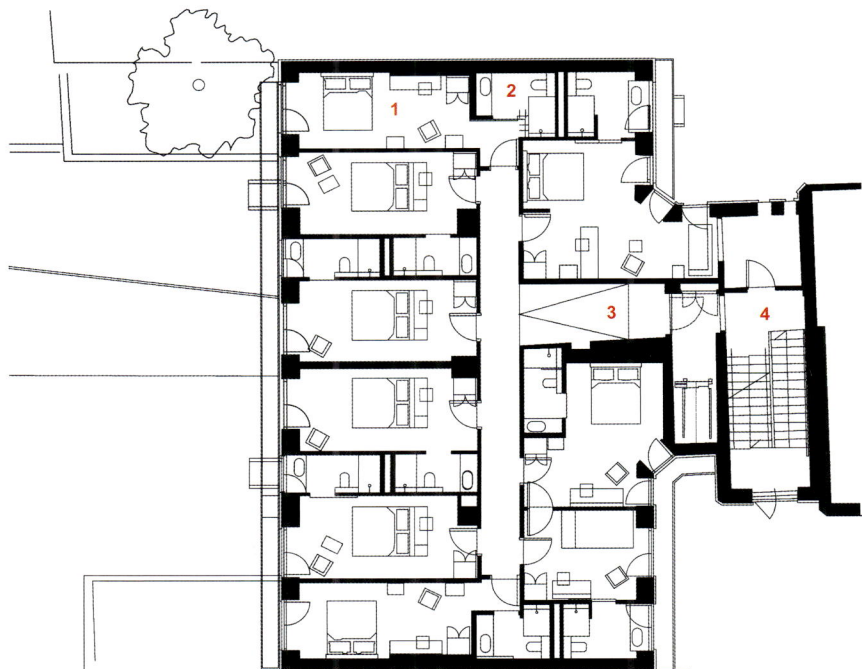

1. Bedroom
2. Washroom
3. Community area
4. Balcony
1. 卧室
2. 卫生间
3. 公共区
4. 阳台

Andel's Lodz

安德尔连锁酒店

Location: Lodz, Poland
Completion: 2009
Designer: Jestico + Whiles
Photographer: Ales Jungmann

地点：波兰 罗兹
翻新时间：2009年
设计：Jestico + Whiles建筑事务所
摄影：亚历克斯·荣格曼

BACKGROUND

Jestico + Whiles was appointed as interior designer, to work alongside executive architects OP Architekten, to create a stunning contemporary hotel from one of the largest Victorian textile factories in Europe. The factory was opened in 1852 by the industrialist Izrael Poznanaski. After WWII it was nationalised and quality and markets declined until it closed in 1997.

SOLUTIONS

The design has been developed in close consultation with the conservation authorities of Lodz and will unlock the true potential of this historic building, of stunning proportions, for the first time in 30 years. The designers' mission was to find an entirely new way for the building to be perceived and experienced, to release the joy and delight locked in the history of the building. This was achieved by contrasting the preserved industrial fabric with strong, but spare, contemporary interventions.

Three stacks of elliptical light wells are punched through the full height of the building. These tapers are aligned obliquely, perceived as unified cones, reaching up through five floors to a glazed rooflight. These "light cones" create a holistic experience at the ground floor and at every bedroom floor, unlocking an ever changing combination of framed views up to the sky and down to the animation of the lobby and lobby bar. Daylight is channelled to illuminate the heart of the deep, open plan ground floor lobby bar and, by night, coloured light stains the ellipses to create towering, kinetic sculptures, or films are projected onto them to create distorted, abstract cinemas, adding discrete animation to a calm space. The soft curves of the light cones contrast with the tight, marching grid of cast iron columns, while the solid, brilliant white balustrades contrast with the softness, texture and warmth of the aged red brick vaulted ceilings.

The hotel includes 180 bedrooms, 80 long stay apartments, a four-level atrium, a stylish bar, café, restaurant with a business centre and a 600-capacity grand conference room. Bedrooms are located to the perimeter of the upper levels, potentially leaving the central bay of each of the deep factory floors deprived of natural light, and over-sized for the required circulation.

Bedrooms are located to the perimeter of the upper levels to take advantage of the high factory windows which let light deep into each floor of the building.

1. Exterior view of the original textile factory
2. Interior staircase of the original textile factory
3. Refurbished exterior and main entrance
1. 原有纺织厂外观景象
2. 原有纺织厂内部楼梯
3. 改造后外观及主入口

4

背景

安德尔酒店由欧洲最大的维多利亚时期纺织工厂建筑改造而成。工厂于1852年由知名工业家Izrael Poznanaski成立，第二次世界大战之后工厂被国家收购，之后随着质量的下降以及市场的衰退，在1997年倒闭，之后建筑一直被荒废。Jestico + Whiles建筑事务所同OP建筑公司合作，旨在将这一废弃的建筑改造成现代风格的酒店。

过程

设计师同罗兹市政府的遗产保护组织紧密合作，借以揭开这座已废弃30多年的历史建筑所蕴含的真正潜力。设计师的任务即为找到一种全新的设计方式，将建筑中隐藏的愉快与喜悦挖掘出来，供前来酒店的客人去发现、去体验。这一想法最终通过建筑已有的结构样式与现代风格的对比而实现。

三个层叠的进光孔贯穿整幢大楼的高度，其呈现椭圆形并在视觉上逐渐变窄。它们斜向排列着，犹如统一规格的圆锥体穿过各个楼层一直延伸到屋顶处。这些"圆锥孔"为一层及上面的客房层带来了不断变幻的视野，同时使得大堂及酒吧活力十足。白天，光线透射进来，照射到一层大堂内；夜晚，彩色的灯光闪烁着，映射出不同的景象，犹如抽象电影一般，为幽静的空间增添了动感气息。柔和的弧形线与整齐排列的铸铁梁柱形成鲜明的对比，现代风格十足的明亮白色栏杆与古老的拱形红砖天花各具特色。

酒店共包括180间客房、80套公寓、四层高的中庭、酒吧、餐厅、可容纳600人的会议室。客房分布在上层的四周，使得中央区域空闲出来，便于光线的照射以及通道的设置。

4. Bar counter and egg-shaped cellarette
5. The soft curves of the light cones contrasting with the tight, marching grid of cast iron columns
6. Reception desk seen from the grid of cast iron columns
7. Lounge area
8. Restaurant

4. 吧台及蛋形的酒柜
5. 柔和的弧形线与整齐排列的铸铁梁柱形成鲜明的对比
6. 从铸铁梁柱间隙望向接待台
7. 休息区
8. 餐厅

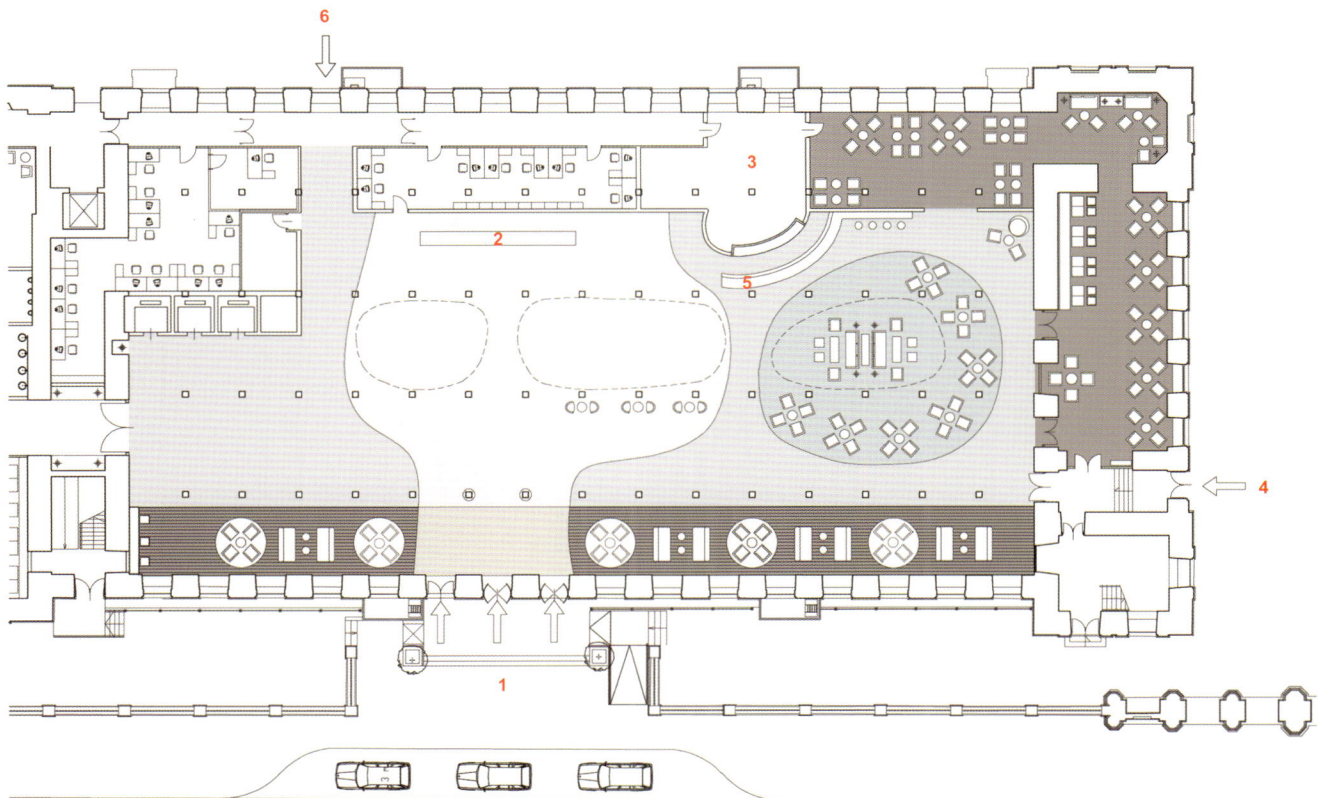

1. Entrance
2. Reception
3. Pantry
4. Café entrance
5. Bar
6. Group entrance

1. 入口
2. 接待台
3. 配餐室
4. 咖啡厅入口
5. 酒吧
6. 团体入口

9. Bedrooms are located to the perimeter of the upper levels to take advantage of the natural light from the high factory windows
10. Three stacks of elliptical light wells punched through the full height of the building
9. 客房分布在上层的四周，使得中央区域空闲出来，便于光线的照射以及通道的设置
10. 三个层叠的进光孔贯穿整幢大楼的高度

11. The distinctive decoration objective echoing the long history of the building
12. The green sofa and chairs as well as the exposed brick wall adding natural feeling to the living room
13. The backdrop to the bed and beddings corresponding each other in colour and pattern
11. 独特的装饰突出了建筑古老的历史
12. 绿色的沙发、椅子以及未经修饰的砖石墙壁增添了自然气息
13. 客房内床的背景墙装饰与床品在色彩及花样上相互呼应

Palafito 1326 Boutique Hotel

Palafito 1326 精品酒店

Location: Castro, Chile
Completion: 2009
Project: Edward Rojas Arquitectos
Designer: Edward Rojas & Juan Fernando Yaez
Photographer: Carlos Mallagaray
Area: 665.14m²

地点：智利 卡斯特罗
翻新时间：2009年
项目设计：艾德华·罗哈斯建筑事务所
设计：艾德华·罗哈斯、胡安·费尔南多·雅奈兹
摄影：卡洛斯·马拉加雷
面积：665.14平方米

BACKGROUND

Gamboa, a pile-dwellings neighbourhood in Castro, capital city of the Great Island of the archipelago of Chiloé, in southern Chile, is a living heritage born from urban marginality at the beginning of the 20th Century. In this century it has consolidated as an icon of wood architecture in Chiloé and of its culture intimately linked with the land and the sea.

The strong urban transformation of the neighbourhood over the past years has made it possible to recycle this unique and precarious constructions made of wood as spaces destined to embrace tourism, what involves a paradigm change, whereas this icon besides being photographed, can now be inhabited by visitors.

OBJECTIVE

The Palafito 1326 is a 12-room boutique hotel, which is part of this process, and seeks with a high level of comfort and great refinement in the use of the essential materials in this culture (wood, fibres and wool), to offer the unique experience of lodging by the shore and enjoy the constantly changing landscape with the rising and falling tides.

SOLUTIONS

The work, in sustainable terms, is proposed as a contemporary re-interpretation of the traditional image of the stilt house, and whereas it is built using the experience of insular carpenters, many of them inhabitants in this neighbourhood, with wood from the native forest and traditional technologies, and incorporating both the recycling of materials from the old building, such as Redwood shingles, doors and windows, and contemporary technologies and materials such as adequate thermal insulators, asphalt roofing, and hermetic double glazed PVC Windows that enable a higher thermal efficiency to the central heating by a great wood boiler chimney, located in the living room.

In contemporary terms and as part of the dialogue between tradition and modernity, the building features as an innovation, a large terrace that makes it possible for guests and visitors to enjoy, above the roof level of the other houses on stilts, of all the spatial, natural and architectural dimension of this particular territory located in the Patagonia.

1,3. Disassembling of the original pile-dwellings
2. Original front view
4. Multi-purpose room and deck on top of the roof, to fully enjoy the landscape, including glass floors to allow light into the hallways.
5. Hotel, sitting on wood stilts in the pile dwelling neighbourhood of Gamboa

1,3. 原有桩住宅拆迁
2. 原有桩住宅正面
4. 屋顶上的多功能房间及露台，可欣赏周围的景观，玻璃地面便于光线照射到下面的通道内
5. 酒店及周围的水上住宅区

6

背景

甘博亚是卡斯特罗（智利南部群岛）著名的桩住宅区，20世纪之初这里被称为"城市边缘的活态遗产"，同时，这里也成为木质建筑及海洋岛屿文化的重要标识。

过去几年里，甘博亚住宅区的大规模改造使得这些独特而又珍贵的木质建筑成为了旅游业的新宠，不仅被用来拍照，同时也被改造成了供游客居住的酒店。

目标

Palafito 1326精品酒店包含12间客房，作为改造工程的一部分，旨在寻求高品质的舒适度、突出地域特色的材质使用，为游客带来独特的居住体验，享受潮汐引起的景观变换。

过程

环保理念作为这一改造项目的主导，将传统的小木屋运用现代化风格诠释。选用的材料以木头为主，主要来自于当地的森林。旧建筑原有的结构，如红木屋顶板、门窗等也被循环利用。现代技术及材质，如隔热板、沥青屋顶、密封双层塑钢玻璃窗等也被引用进来。

设计中一直寻求传统与现代的对话，而这一建筑则是作为整个对话的一部分。屋顶上宽敞的露台别具特色，为客人提供了欣赏周围自然景观、空间构造及建筑特色的完美场所。

6. Refurbished exterior view
7. Exterior staircase, hallways and terraces, for both escape and appreciation of the surroundings
8. Reception lobby, with front desk covered with alerce recycled shingles from the old house
9. Hallway facing the landscape
10. Dining room looking out onto the terrace and the seashore
11. Second floor hallway facing landscape and access to internet room

6. 改造后外观
7. 室外楼梯、通道及露台用于紧急情况逃生及欣赏周围的景色
8. 大堂，接待台台面由从旧屋中回收的木板制成
9. 朝向室外景观区的通道
10. 餐厅朝向室外露台以及海岸
11. 二层朝向室外景观区的通道以及通往网络室的入口

1. Access foyer
2. Reception
3. Staircase and hallways
4. Office
5. Laundry
6. Bedroom
7. Living - dining room - kitchen
8. Wood storage
9. Terrace
10. Exterior staircase and hallways

1. 入口大厅
2. 接待台
3. 楼梯及走廊
4. 办公区
5. 洗衣房
6. 卧室
7. 起居室——餐厅——厨房
8. 木材存储区
9. 露台
10. 室外楼梯及走廊

12. Living room, with fireplace furnace and benches covered with recycled alerce shingles
13. Bedroom with wood panelling, framing the landscape
14. Bedroom viewing the neighbourhood, with hand-crafted furniture and textiles
12. 客厅，壁炉及长椅采用旧屋中回收的木板装饰
13. 朝向自然景观区一侧的卧室
14. 朝向住宅区一侧的卧室，手工制作的家具及织物特色十足

13

14

St Giles Hotel

圣·基拉斯酒店

Location: Edinburgh, Scotland
Completion: August 2009
Designer: Holmes Partnership
Photographer: Holmes Partnership
Area: 4,000m²
地点：苏格兰 爱丁堡
完成时间：2009年8月
设计：霍姆斯合伙人事务所
摄影：霍姆斯合伙人事务所
面积：4000平方米

BACKGROUND

Designed by David Bryce in 1870, the buildings at 12-26 St Giles Street were originally built as newspaper offices and printing workshops, later converted for a variety of uses and used latterly as offices for Edinburgh City. Four storeys above ground level at St Giles Street, the site falls away steeply to the north to Market Street and the Waverley Valley to Princes St and beyond.

OBJECTIVE

The properties were converted to an apartment hotel with a range of accommodation, from studio and one-bedroom apartments to suites together with all the ancillary accommodation expected of high quality accommodation of this kind.

SOLUTIONS

Extensive structural and fabric repairs were undertaken. The interior features which include panelled walls and decorative ceilings were fully refurbished.

Many rooms have spectacular views over Princes Street and the French contemporary décor of each apartment incorporates high specification fitments and finishes with fully tiled bathrooms and fully fitted kitchens in every bedroom including the four accessible apartments.

Located only yards from the Royal Mile the new hotel provides high quality self catering accommodation for both business and leisure visitors to the centre of Edinburgh.

1. Facade of the original building
2-3. Offices inside the original building
4. Lobby - the black table and bookshelf going well with each other, while the white chair and black chair forming a strong visual contrast
1. 原有建筑外观
2-3. 原有建筑内办公室
4. 大堂一角，黑色的桌子与黑色的书架交相呼应，而黑白色彩鲜明的座椅则在视觉上形成强烈的对比

背景

酒店原有建筑位于圣·基拉斯大街12-26号,由大卫·布赖斯(David Bryce)于1870年设计而成。最初被用作报社及印刷工厂,随后经过多次改造,最近一次改造成爱丁堡市政府办事处。该建筑地上共为四层,其所处地块地势略微向北侧的市场大街方向倾斜。

目标

此次改造的目标是将其打造成一个公寓式酒店,包括工作室、单人房及套房等,其所有的辅助设施也都要满足高标准要求。

过程

原有结构被进行大规模地修复,室内空间特色元素如镶板墙壁及装饰天花等全部经过重新处理。

大多数房间内可以观赏到王子街(Princes Street)的美丽景致,现代法式装饰风格洋溢在每个房间内,完全由瓷砖打造的浴室以及设备齐全的厨房更是别具特色。

酒店距离皇家麦尔大道(Royal Mile)仅有几步之遥,旨在为商务人士及旅行者们提供高质量的住宿服务。

5. Reception desk - three illuminating fixtures dangling from the ceiling throwing a pool of yellow light
6. Lobby lounge - the contrast of black and white being stressed here
5. 接待台,三个灯饰从天花上垂悬下来,洒下一抹浅黄的光线
6. 大堂休息区,黑白色彩的对比格外突出

1. Main Entrance
2. Reception
3. Luggage
4. Office
5. Lift
6. Studio
7. Accessible studio
8. Restaurant
9. Studio with table
10. Room Dining / Meeting
11. Bed

1. 主入口
2. 接待台
3. 行李处
4. 办公区
5. 电梯
6. 商务客房
7. 无障碍入口商务客房
8. 餐厅
9. 商务客房（带小型会议区）
10. 餐厅／会议室
11. 卧室

7. The stripe patterns on the wall injecting livliness and playfulness into the whole space and the framed paintings on the wall being extremely eye-catching
8. The bedroom giving off touches of cosiness by the application of warm colours
9. The detailed design on the window bringing interest to the guestroom and the light blue creating a fresh enviroment

7. 墙壁上条纹图案为整个空间注入了活力，画框装饰则更加吸引眼球
8. 暖色调的运用赋予整个卧室温馨的氛围
9. 窗户上的细节设计增添趣味性，淡蓝色调的选择营造清新的环境

1. Studio
2. Stair
3. Suite
4. Lift
5. Deluxe studio
1. 商务套房
2. 楼梯
3. 套房
4. 电梯
5. 双人商务套房

Wald-und Schlosshotel Friedrichsruhe

瓦尔德酒店

Location: Friedrichsruhe, Germany
Completion: November 2008
Designer: Niki Szilagyi Interior Architecture
地点：德国 弗里德里希斯鲁厄
翻新时间：2008年11月
设计：Niki Szilagyi 室内建筑公司

OBJECTIVE

The task was to integrate some components of the castle and the existing hotel into the new concept. The design is modern but not obtrusive.

SOLUTIONS

The first highlight upon, entering the lobby, is the reception desk made of makassar wood with a unique frieze in the front made by the painter Johannes Klinger. With its ornamentation in gold it reflects the flair of this landscape. The floor lands the room a harmonious structure - it is made of French lime stone with cognac-coloured inlay work made of oak. With an area of 400 m² the lobby with its lighting concept and flexible furniture is the perfect place for exhibitions or events. It is subtly separated into two areas: one in front of the restaurant with a little suite and a second in front of the reception. The second is comprising a mobile shop,which can be moved into the near-by stockroom. Close to the reception you find the dressing room as the entrance to the whole spa-area.

The design of the restaurant compliments the menu. The open plan kitchen is inspired by typical Italian kitchens, and conceived by a specialist.

The entrance to the spa next to the reception is kept narrow to surprise the guest with a gorgeous view of the spacious pool area.

The swimming hall with gallery extends to the first floor and is flooded with daylight - due to the roof light.

Different themes are shown in the saunas and steam baths. An imposing feature serves on the one hand as a towel shelf - on the other hand as an illuminated partition. Adjacent to which you will find a water bar with silver quartzite flagstones. A booth consisting of a steam bath, sauna, showers and space for relaxation is located at the back of this area. It can be used exclusively for a group of women, for example.

An onyx bar is the centrepiece of the bistro. Thanks to its height, this bar is very comfortable for the guest. Here he can sit and eat in his bathrobe, take a quick snack at a coffee table or the intimacy of one of the niches, as in a classical bar.

A large-scale wine fridge filled with carefully selected wines forms the background of the bar. Ordering is easy as the staff is close at hand to provide meals cooked in the restaurant. The floor made of solid oak combined with French lime stone and teak-furniture surrounded by a cosy collection of natural fabrics in sunny colours - create an atmosphere of luxurious relaxation.

Big windows allow light to flood the room, and enable the guest to move outside where he can lounge in warm summer months. The colonnade connects the hotel and the new spa. It starts with a corridor. A change of materials - from soft French carpet to French lime malm brick - together with the aromatic herb-garden - pieces of modern art allow total relaxation. Arriving at the entrance of the new spa the guest is introduced to an outstanding adventure in stress relief. Custom-designed carpets cover the floors of the rooms and corridors creating a warm and cosy atmosphere in combination with the walnut parquet. The bathrooms are flooded with daylight that you can shade with venetian blinds. The suites on the ground floor are appointed with a patio. The textile design inside the rooms goes hand in hand with the vegetation in the garden - again a harmonic dialogue between inside and outside.

1. Lobby before expansion of the original hotel
2. Spa before expansion of the original hotel
3. Guestroom before expansion of the original hotel
4. The existing castle and outdoor pool
1. 酒店扩建之前大堂
2. 酒店扩建之前水疗馆
3. 酒店扩建之前客房
4. 酒店扩建部分城堡及室外泳池

4

1. Atrium
2. Indoor pool
3. Wandelgang
1. 中庭
2. 室内泳池
3. 走廊

5

目标

设计的任务是将（原有的）城堡式酒店的一部分和现在（扩建）的酒店融合为一体，形成新的设计理念。酒店的风格现代且不浮华。

过程

走入大堂，首先映入眼帘的是望加锡木做的接待台，其正面独特的毛绒画是艺术家Johannes Klinger的作品，金色镶边的图画回应了周围的景观。地面铺着法国石灰石地面和上等科捏克白兰地颜色（金黄色）的橡木，和室内的装饰很谐调。400平方米的大堂里，灯光和谐，家具可以灵活地移动，是个展览或集会的好地方。这里还细分为两部分：一部分是餐厅前面的一个小套间，另一部分是接待台前面的区域，这里有一个可移动的商店，它能被移动到附近的衣帽间处。接待台附近的一间梳妆室也是整个温泉区的入口。

餐厅的设计很好地回应了这里供应的菜肴。开敞的厨房是典型的意大利式厨房，由专业人士设计。

接待台旁边进入温泉区的入口狭窄，它和华丽、宽敞的泳池区形成对比，让客人大为惊喜。

游泳大厅和休息平台跨越两层空间，因为设有采光屋顶，这里阳光充足。

桑拿区和蒸汽浴室有不同的主题。一个让人印象深刻的亮点是两边即做毛巾架又做灯光隔断的设计。临接这里有一个水吧，用银白色的石英岩铺地。其后面的一小隔间里有蒸汽浴、桑拿、淋浴和休息空间，也可以单独为女人们所用。

玛瑙材料的吧台。它的高度适中，让客人感觉很舒服。客人可以穿着睡衣在这里吃东西、喝酒，在咖啡台处或在一个私密的角落里吃快餐，就像在典型的酒吧里那样。

大型的葡萄酒冰箱内放着精选的葡萄酒，成为了酒吧的背景（墙）。在这里点菜是很方便的，服务员就在周围，等待给客人上菜。地面铺着硬橡木和法国的石灰石，柚木家具被环绕在颜色明快、舒适的天然织物中，创造了一个豪华的休闲氛围。

大面积的（落地）玻璃窗让阳光洒入室内，客人可以在温暖的夏日里搬到室外休闲。酒店和新的温泉区用柱廊连接。一段走廊是它的开端。变换的材料：从柔软的法国地毯到法国石灰砂砖过渡到芬芳的香草花园，还有摆放的几件现代艺术品让客人完全放松。到了新温泉区的入口，客人（好似）来到了一个出色的奇境中来，（很快）解除了压力。铺在房间内和走廊的地毯是定做的，它和核桃木地板一起创造了一个温馨、舒适的环境。浴室的阳光充足，也可以用威尼斯遮帘挡视线。一层的套房有院子。室内的织物设计和花园内的植物相呼应，这种设计语汇使室内外和谐相连。

5. Refurbished lobby reception
6. Restaurant
7. Spa bistro bar
5. 改造后大堂及接待台
6. 餐厅
7. 水疗馆自助餐吧

6

7

8

8. An imposing feature serving on the one hand as a towel shelf and on the other hand as an illuminated partition in the sauna area
9. Double treatment room with the central part of the ceiling made by patterned glass
10. Colours in the spa changing from light yellow to deep orange creating a warm and cosy atmosphere
8. 桑拿区的特色既为及用作毛巾柜又用作灯饰隔断的橱柜
9. 双人理疗室天花中央部分由印花玻璃打造
10. 水疗区内的色彩从浅黄到深橘色，营造了温馨舒适的氛围

11

11. Spa area in the presidential suite
12. Wooden furniture, floral patterned carpet as well as yellow light bringing naturalness and warmth to the deluxe suite
13. The partition wall between living room and bedroom of the deluxe suite performing as fireplace and book shelf
11. 总统套房内的水疗区
12. 双人豪华套房内，木家具、花朵图案地毯以及黄色灯光使空间增添了自然气息与暖意
13. 豪华套房内，客厅及卧室之间的隔断墙同时兼壁炉和书架的功能

Hotel Denit

德尼特酒店

Location: Barcelona, Spain
Completion: 2008
Designer: GCA Arquitectes Associats
(Director of the project: Josep Riu, Interior designer: Beatriz Cosials, Architect: Anna Rey)
Photographer: Hotel Denit
Area: 2,500m²
地点：西班牙 巴塞罗那
翻新时间：2008 年
设计：GCA建筑事务所
（项目总监：约瑟普·秋，室内设计：比阿特丽斯·科斯阿尔斯，建筑设计：安娜·雷伊）
摄影：德尼特酒店
面积：2500平方米

BACKGROUND

The project consists of the integral rehabilitation of an apartment building of the 1950s, placed in the historical centre of the city of Barcelona, in a journeyed and pedestrian street, next to the commercial zone of Portal de l'Angel and the Catalunya Square. The building is formed by cellar, ground floor, four floors and attic with a court on the central part of the frontage to street.

OBJECTIVE

The new use is to design a 3-star hotel. The new functional programme is organised of the following form: on the four-type floors are five double rooms, two individual rooms and one office. In the attic are five double rooms, one individual and one office.

SOLUTIONS

The new use implies having two stairs, a clients' elevator and a hoist of service; this nucleus is located in the centre of the plant.

In all the noble zones of the hotel it has placed a continued floor made of resins of white colour Ral 9010 with chips grey incorporated. The walls in white, as well as the roofs make themselves up to the plastic in the same colour. Vertical elements as the elevators in all the plants, and exempt pieces as the box of the hall or the kiosk of reception, are constructed in plate of black varnished colourless dull iron.

The general lighting, with the exception of decorative lights, answers to the concept of "low cost" since they are fluorescence leaned in hollows realised in false roofs, remaining completely visible.

In an axial way, the reception (understood as multifunctional element of kiosk, bar, reception) is the essential part of the space and shares space with two areas of rest, one of lounge and other one of community table for different and Internet uses, both areas inside a game of walls and roofs with hollows that hide the indirect lights. The reception treated as the furniture multiuse with functions of clients' reception, shop - kiosk, bookstore, selling of aperitifs, and information of the city. The above mentioned furniture helps to organise the circulation in the floor, gives certain privacy to the vertical accesses, elevators and stairs, presiding at the space.

This lounge organises about a community table realised in oak, two built-in banks placed in the longitudinal sense made by vertical cylinders upholstered in white polyskin. All the pieces of seat are solved by the same model in chair or in stool; it is the model Smile de Andreu World.

This space closes with a few big sliding white doors that integrated into the walls when these are opened. They separate the space "breakfast" from the foyer of arrival to plant in the one that finds the bar of the office and the access to the public services.

The hotel is understood as a construction in white, inside which there are special pieces realised in wood of oak or in black iron, but the second colour appears inside the building; this is the colour "denit", one somewhat violet blue, with that there are several solved elements as, for example, the signs or the colours of certain false roofs and decorative pieces. In the corridors of the floors of rooms, and although there is supported the concept raised in the noble zones, the principal criterion is that of acoustic and functional comfort. The rooms answer to the criteria of simplicity and functionality joined the concept of modernity. The unit of the room is understood as a white box, with walls and identical roofs, and the paving is the same that it continued of the rest of the hotel.

1. Entrance of the original apartment building
2. Façade of the original apartment building
3. Corridor
4. Lobby
1. 原有公寓建筑入口
2. 原有公寓建筑外观
3. 走廊
4. 大堂

dreit

Night&breakfast'bcn

4

5

背景

酒店由位于巴塞罗那市中心区的一幢建于20世纪50年代的公寓改建而成。原有建筑坐落在一条狭窄的街道内，与Portal de l'Angel商业区及加泰罗尼亚广场相邻，由地下酒窖、一层公共区、四层客房、阁楼以及朝向街道一侧的小庭院构成。

目标

设计的目标是打造一个新的三星酒店，其功能如下：客房层每层包括5间双人套房、两间单人房及办公区；阁楼包括5间双人套房、一间单人房及办公区。

过程

建筑的新功能意味着其要增添两部电梯，一部专为客人使用，另一部为酒店服务人员使用，并设置在中央区域。

酒店所有区域内地面全部采用白色树脂铺设，其中掺杂着点点的灰色。墙面及天花同样采用白色粉饰，打造一个梦幻的世界。电梯、接待台处的柜子以及电梯井采用黑色铁板打造，与空间整体氛围形成对比。

除一些装饰性灯饰，空间整体照明以"低成本"的理念为主，荧光灯排列在假屋顶形成的空隙内，完全可见。

大堂中轴区内，接待台成为了中心元素，两侧分别为休息区及供客人上网使用的大桌子。此外，这两个区域内的墙壁及天花内"隐藏"着间接照明设备。接待处本身就如同一个多功能的家具，既是客人登记台，又是一个小商店，供应书籍或开胃酒等，同时更是信息栏，提供城市信息。这一"家具"也是空间通道的标识，将电梯及楼梯等"藏"在身后。

休息区内的大桌子采用橡木打造，嵌入墙壁的圆柱形结构与桌面平行并悬垂在其上面，采用白色粉饰。所有的座椅在形状上完全统一，连续感十足。休息区与一面嵌在墙壁中的白色滑动门相邻，将早餐区同大厅分离开来，同时通往酒吧及公共区。酒店装饰以白色为主，仅有少数的饰品由橡木及黑铁打造。而另外的颜色即为酒店标志色——紫蓝色，体现在酒店标识、假天花上。客房走廊设计除遵循公共区设计理念之外，更注重隔音效果。客房彰显简约性、功能性及现代性。所有的客房犹如一个白色的大盒子，同样的墙壁、同样的屋顶、同样的地面。

5. Lounge area
6. Office area
5. 休息区
6. 大堂办公区

1. Lobby   1. 大堂
2. Reception   2. 接待台
3. Shop-library   3. 商店及图书室
4. Room   4. 客房
5. Interior courtyard 5. 室内庭院
6. Private area   6. 私人休息区
7. Service access 7. 服务区入口

7-8. Dining area
9. Guestroom
7-8. 就餐区
9. 客房

Eynsham Hall

茵斯汉姆酒店

Location: Oxford, UK
Completion: 2008
Designer: Project Orange
Photographer: Richard Learoyd
地点：英国 牛津
翻新时间：2008年
设计：Project Orange 设计事务所
摄影：理查德·利罗伊德

BACKGROUND

Situated on the outskirts of the city of Oxford in extensive grounds, Eynsham Hall is a magnificent countryside mansion. Completely remodelled in 1908 by Lady Mason, Eynsham Hall is wonderful example of the Jacobean Style very much in vogue at the turn of the twentieth century.

OBJECTIVE

Having suffered from various interim modifications while in use as a conference centre, the property is about to undergo an ambitious programme of works and once again will be transformed into a luxurious, country house hotel.

SOLUTIONS

Project Orange were invited to redesign the hotel bar in order to test design strategies and methods for modifying a historic building such as Eynsham. Originally the gunroom for the Mason family's collection of firearms, the new bar, now known as The Gunroom, fuses old with new, with a touch of art deco glamour.

The bar counter is an island of decadent, black glass and polished chrome sitting in the gigantic, leaded bay window overlooking the grounds. The impressively high ceilings have even helped to find a home for some of the estate's vast collection of antlers.

The challenge was to convince the client that they could insert modern furniture and quirky pieces into this traditional setting to create a wow factor. Working along side Studio Myerscough they came up with the narrative of the "hunt" thinking about people in their traditional yellow riding jackets with brass buttons and the gloss of their polished boots. This seemed to situate the project in its history allowing us to maximise the potential of this listed room.

The bedrooms and bathrooms:
With no two bedrooms alike on the first floor of the hall and with a series of very tight bedroom layouts to the second floor, the designers were presented with a very difficult design challenge. The bathroom is tiled in dramatic black marble taking its clue from the fireplace.

Meeting room:
Formerly the private office of Lord Mason, an enormous leaded window with views out into the extensive grounds. An interconnecting door with the meeting room's neighbour the Gun Room bar, also introduced a further use as a potential private dining space.

1. Meeting room of the original countryside mansion
2. Bedroom of the original countryside mansion
3. Bar of the original countryside mansion
4. Front view of the hotel
5. Main entrance and façade detail
1. 原有乡村公馆会议室
2. 原有乡村公馆卧室
3. 原有乡村公馆酒吧
4. 酒店正面
5. 主入口及外观细节

背景

酒店原为一幢宏伟的乡村公馆，坐落在牛津城郊。1908年由Lady Mason操刀将其改造成雅克布风格建筑，并使其成为20世纪之初极为流行的建筑样式。

目标

酒店的内部历经多次改造，此次则为进行更大规模的改建：从会议中心再次改建成豪华的、乡村别墅型酒店。

过程

酒吧的改造用以检验改造像茵斯汉姆这样一座历史性建筑的设计方法和策略。梅森家族收藏枪支的房间被用做新的酒吧间，并将其命名为"军械库"。设计中将古老与现代风格结合起来，艺术品起到点睛作用。原来使用的橡木板被修整一新。吧台采用黑色玻璃及剖光铬材质打造，摆放在凸窗前，营造出颓废风格。高挑的顶棚给人以深刻印象，并为庄园主收藏的鹿角提供了"居住空间"。

要说服客户同意在传统的背景里摆放一些现代的家具和稀奇的器物，并能给人眼前一亮的感觉，是有些难度的。设计师同Myerscough事务所仔细商讨，最终构思出"狩猎"的主题，勾画了人们在穿着缝有黄铜扣的黄色骑士上衣和擦得亮亮的皮靴而打猎的场景。这样的设计就符合项目本身的历史文脉，使这个历史保护区的特色被充分展示。

客房与洗手间：

二层，酒店的豪华客房布置各不相同；三层，客房布局非常紧凑，因此为设计改造工作带来极大的挑战。洗手间采用浪漫的黑色大理石贴面，与壁炉的颜色相呼应。

会议室：

由原来的梅森勋爵的私人办公室改造成会议室，三面设计着窗户，将庭院内的景色尽收眼底。会议室同酒吧之间通过一扇门隔开，当然这里还可用作私人宴会厅。

6. The high ceiling boasting classical style and the traditional-patterned windows in the lobby "inviting" outdoor landscape inside
7. The meeting room being redesigned from the private office of Lord Mason
6. 大堂内，高挑的顶篷突出古典特色，古老风格的窗户将户外的景观引入进来
7. 会议室由梅森勋爵的私人办公室改造而来

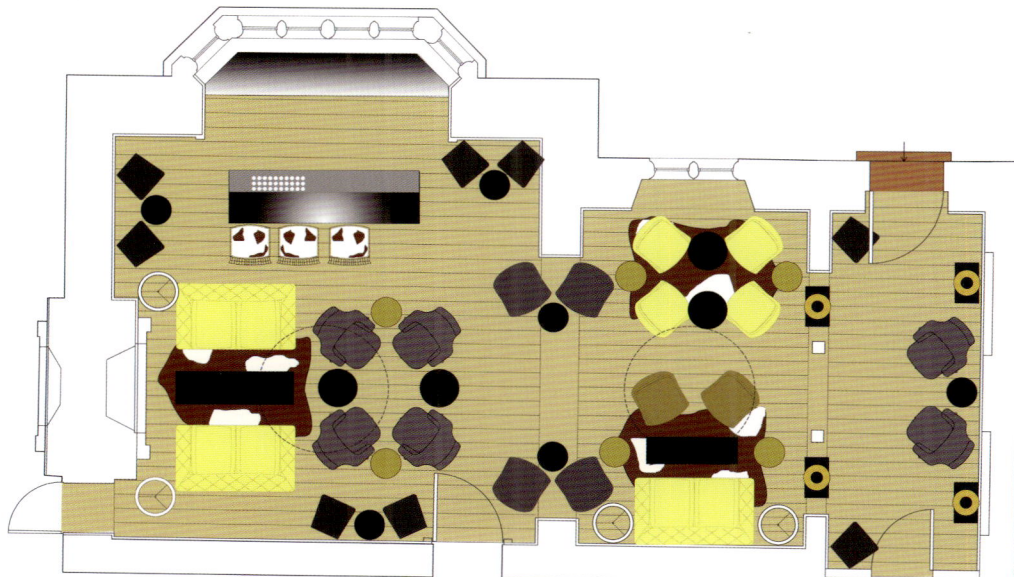

8. Bar neighbouring the meeting room remodelled from the original gunroom
9. The pure coloured wall and ceiling providing a clean background for furnishings and the fireplace as well as fur carpet creating a warm, homey atmosphere in the guestroom
10. The wall being tiled in dramatic black marble to form a striking contrast with white washbasin and bathtub

8. 酒吧由原来的军械库改造而来
9. 客房内，纯色的墙壁及天花打造了一个简约整洁的背景，壁炉及皮毛地毯则营造了温馨的居家氛围
10. 卫生间内，黑色大理石墙壁与白色舆洗盆及浴缸形成鲜明对比

Ca' Sagredo Hotel

卡撒乐度酒店

Location: Venice, Italy
Completion: 2007
Designer: Academy of Fine Arts of Venice
Photographer: Wladimiro Speranzoni
Area: 6,000m²

地点：意大利 威尼斯
翻新时间：2007年
设计：威尼斯美术学院
摄影：瓦拉迪米罗·斯皮兰佐尼
面积：6000平方米

BACKGROUND

This palace was originally owned by the Morosini family and was purchased at the start of the 18th century by the Sagredos, a noble family who had lived in the Santa Sofia district for centuries. The façade onto the Grand Canal is proof of the Byzantine origin of the building, which was altered several times in subsequent centuries. The original ground floor with the doors leading onto the water and the first floor with its tall arch windows topped on slim pillars, were completed in the 15th century by the addition of second floor, which has tracery frieze around the middle mullioned windows of the Portego or central hall. Despite centuries and changes the Palace still preserves its untouched beauty.

OBJECTIVE

This is a real treasure: the restoration is to carefully enhance the original architectural features, delighting the visitor while evoking respect for times gone by.

SOLUTIONS

Music ballroom:
In the golden and precious Music Ballroom, numerous frescoes attributed to Gaspare Diziani completely cover the walls and ceiling. Along the walls stand, inside sham niches, the monochromatic figures of Minerva, Neptune, Cibel, Mars, Venus, Mercury, Juno and Jupiter. Splendid chandeliers in gold leaf hang from the ceiling, and the floor is embellished with the coat of arms of the Sagredo family. The frescos on one wall act as a camouflage for a door to the secret passage which once led to the "Casino Sagredo". This passageway was used by mistresses during balls to discretely make their ways to the master's alcove. The ballroom is particularly interesting: it extends to a height of two floors.

Suite:
Many of the rooms, a number of which lead off from the portego, are decorated with very refined and elegant stucco work, probably by Abbondio Stazio and Carpoforo Mazzetti Tecalla, who were also responsible for decorating the place's mezzanine floor.

Staircase:
The staircase, designed by the architect Andrea Tirali in the third decade of the 18th century and completed in 1732 when Pietro Longhi started to paint the frescoes which still surround it, was just a part of an overall renovation projected for the place, undertaken by Gerardo. Two marble cherubs by Francesco Bertos decorate the entrance to the staircase, glancing to the incoming guests. The floor is in mosaic, decorated with elegant coloured volutes.

1. Old Music Ballroom
2. Old Portego Hall
3. Refurbished exterior view
4. View from afar
1. 原有宫殿内音乐舞厅
2. 原有宫殿内入口大厅
3. 改造后外观
4. 远景

背景

卡撒乐度酒店原来是一座威尼斯风格宫殿，位于大运河沿岸，最初为莫罗西尼家族拥有。18世纪初期，卡撒乐度家族将其买下，酒店的名字即源于此。建筑朝向大运河一侧的外观呈现出拜占庭建筑风格，原有的一层一直通往岸边，二层带有采用细柱支撑的高大拱形窗户。15世纪建筑加盖了一层，窗户带有花格装饰。然而，即使随着时光的变迁，它所呈现的美感却一直未变。

目标

建筑本身便是真正的宝物，其翻新工作旨在突出原有的建筑特色，令游客对其感到愉悦。即使随着时间的流逝，它给人那种肃然起敬的感觉也不会随之减少。

音乐舞厅：

舞厅可用"金碧辉煌"来形容，墙壁及天花上画满精美的壁画。墙壁上画着许多神话人物肖像，包括智慧、技术和发明女神米纳瓦、海神尼普顿、美神维纳斯等。金箔吊灯从天花上悬垂下来，地面采用卡撒乐度家族的盾徽修饰。另外，墙壁上的壁画后面隐藏着通往密道的大门。

过程

套房：

酒店中的大部分套房全部沿着portego展开，并采用精致典雅的灰泥材质装饰。

楼梯：

原有楼梯由建筑师安德里亚·蒂拉里（Andrea Tirali）于18世纪30年代设计而成，并于1732年竣工。正是在那个时候，画家皮埃特罗·隆吉（Pietro Longhi）开始绘制楼梯周围的壁画，并一直保留到现在。楼梯作为整个改建工程的一部分，由格拉多负责。采用大理石材质打造的两个"小天使"摆放在入口，欢迎着到来的客人。地面呈现出彩色马赛克图案造型，别具特色。

5. Portego Hall
6. Music Ballroom with numerous frescos covering wall and ceiling
7. Detail view of the ceiling
5. 改建后入口大厅
6. 音乐舞厅内墙壁及天花板上布满了壁画
7. 天花上的细节设计

8. Lobby staircase
9. Detail of the Portego Hall
10. Ceiling of library suite
11. High ceiling of library suite
12. Corner view of library

8. 大堂楼梯
9. 入口大厅细节设计
10. 带有图书室的套房天花
11. 带有图书室的套房高挑的顶篷
12. 套房一角

1. Music hall
2. Tiepolo hall
3. Doge hall
4. Amigoni hall
5. Portego
6. Suite
7. Lift
8. Staircase

1. 音乐厅
2. "帝保罗"舞厅
3. "公爵"舞厅
4. "阿米格尼"舞厅
5. 入口大厅
6. 套房
7. 电梯间
8. 楼梯

The Park Hotel Mumbai

公园酒店

Location: Mumbai, India
Completion: 2007
Designer: Project Orange
Photographer: Amit Pasricha
Area: 6,700m²
地点：印度 孟买
翻新时间：2007年
设计：Project Orange 设计事务所
摄影：阿米特·帕斯里查
面积：6700平方米

OBJECTIVE

The building had been abandoned for 20 years never having been completed. Project Orange was asked to come up with an architectural and interior strategy to complete the building, create a series of gardens and design a poolside area. The concept behind the latest Park Hotel was to create an iconic hotel for New Mumbai.

SOLUTIONS

They went back to original Modernist intentions of the building and sought to tidy up the exterior and to render the whole building white. This may seem impractical, and yet whitewashing is a tradition in this area and is done each year after the Monsoon. A double glazed window system was installed and a roof was designed to enclose the atrium, allowing the space to be protected. By night the elevation is illuminated by low wattage amber and blue LED lights. The poolside terrace is characterised by a huge abstract glass mosaic pattern in blues, greys and oranges creating a graphic backdrop for the dramatic black elliptical pool.

The brief for the public areas was to create a four stars plus hotel with flexibility allowing weekday trade to convert to weekend tourist stays. They therefore designed two free-form pods on the ground floor containing the bar, the snug and back office. The orange plastered bar breaks out to address the terrace, while the reception/snug is clad in hand-made plaster jali work addressing the lobby and coffee shop. The bedrooms are generously proportioned with high ceilings, decorated in a classic neutral palate lifted by flashes of bright colour. The bathrooms all have huge showers and a teak vanity counter with the length of the room, which is clad in pristine white tiles. The wardrobe is a one-stop-shop for the guest containing hanging space, a safe, a fridge and a tea tray.

The interiors mix traditional Indian patterns and textures with clean lines and modern shapes, bringing together East and West. The facilities include reception, bar, coffee shop and poolside area on the ground floor as well as the "Bamboo" restaurant seating up to 60 persons. On the upper floors there is a large banqueting hall adjacent to three interconnecting business suites. The bedrooms are arranged around an enclosed atrium, bringing natural light into the circulation areas. On the top floor are larger rooms with a larder facility for long stay guests; there is one master suite, a gym and spa.

In addition, Project Orange designed all the furniture and most of the fabrics for this project. The hotel was listed in the top 10 Wallpaper Business Hotels 2008.

Structure:
80 rooms, 1 restaurant, 1 café, 1 bar, 1 shop, 1 spa, 1 gym, 1 swimming pool, 1 conference room, 1 banqueting room and 1 garden.

1. Exterior view of the original building
2. Original outdoor pool
3. Atrium of the original building
4. Refurbished exterior and outdoor pool
1. 原有建筑外观
2. 原有泳池
3. 原有建筑庭院
4. 改造之后外观及室外泳池

目标

原有建筑一直没有竣工，并荒废了20年。Project Orange 设计事务所负责建筑外观及室内设计，同时打造室外花园及泳池四周区域的规划，目标即为在孟买地区打造一个标志性酒店。

过程

设计师试图还原建筑的最初用途，将外观清理干净之后饰以纯白色彩。这似乎不切实际，但刷白是这一地区的传统，每年季风过后，建筑外观都会重新粉饰。新安的双层玻璃窗以及中庭屋顶旨在将室内空间更好的保护，夜晚外观在蓝色LED灯的照射下格外美丽。泳池边的露台采用蓝、灰、黄三色玻璃马赛克图案装饰，别具特色，同时为泳池营造了一个完美背景。

酒店内公共空间要求达到超四星级标准，满足商务人士需求的同时为观光旅游的客人提供完美的休息之所。设计师在一层打造了两个自由形态的"小岛"，包括酒吧、接待台/休息区及后台办公区。其中，酒吧内以橘黄色调为主，一直延展到室外露台处，接待台/休息区采用当地手工制造灰泥装饰，将大堂及咖啡厅突显出来。客房内采用中性色调装饰，偶尔几抹亮色格外显眼，高大的天花则带来开阔感。浴室内安装有淋浴，柚木梳妆台横跨整个空间。衣橱犹如一站式商店，立面设计着悬挂空间、保险柜、冰箱及茶盘等。

室内装饰融合印度传统图案、线条清晰的织品以及现代风格十足的装饰品，将东、西方特色完美搭配。一层主要包括接待台、酒吧、咖啡厅及"竹子"餐厅（可容纳60人），上层则包括宴会厅以及商务套房等。客房环绕着中庭排列，将自然光线引入进来。顶层客房面积较大，带有食品室、水疗室及健身馆等，主要为长期居住的客人准备。

此外，酒店内所有家具及大部分装饰结构均由Project Orange设计事务所亲自打造，该酒店更于2008年荣获"十大商务酒店"的称号。

酒店空间里，包括80间客房、一个餐厅、一个咖啡厅、一个酒吧、一间商店、水疗室、健身馆、游泳池、会议室、宴会厅及花园。

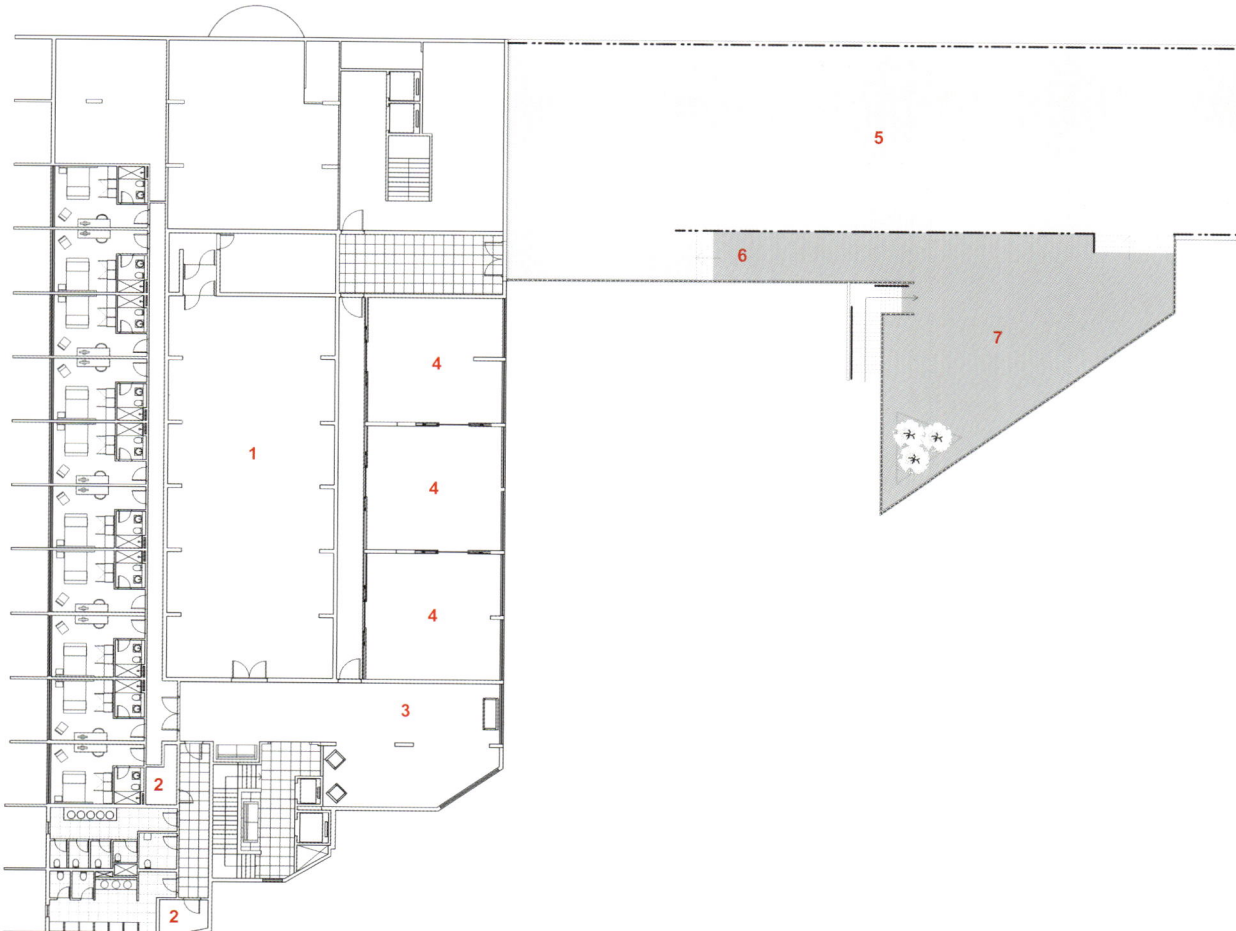

1. Main banqueting
2. Store
3. Pre-function space
4. Function room
5. Banqueting terrace
6. Exit to terrace
7. Roof garden
1. 主宴会厅
2. 储物间
3. 宴会厅接待区
4. 功能区
5. 露台宴会厅
6. 露台出口
7. 屋顶花园

Chiswick Moran Hotel

克里斯莫兰酒店

Location: London, UK
Completion: 2006
Architecture: Capital Architecture
Interior Designer: Project Orange
Photographer: Gareth Gardner
Area: 7,330m²
地点：英国 伦敦
翻新时间：2006年
建筑设计：Capital 建筑事务所
室内设计：Project Orange 设计事务所
摄影：加雷斯·加德纳
面积：7330平方米

BACKGROUND

This 120-bedroom hotel is situated on Chiswick High Road in a vibrant quarter of south west London, housed in a 1960s' former office building.

OBJECTIVE

The main concept for the hotel is "West Coast/ West London". The vibe is a contemporary evocation of 1960s' California, a link between this happening area of London and the glamour of LA.

SOLUTIONS

The hotel is announced by tropical palms and a sweeping porte-cochere, beneath a cantilevered concrete canopy. A palette of heavily veined marble and stained oak boarding unites a flowing sequence of lobby, bar and restaurant. Above the double height lobby is a vast bespoke polished steel and plexi-glass chandelier.

The Globe Bar takes its name from the huge shimmering globe that pivots between reception and bar defined by its fifteen-metre counter in marble and burnt orange leather couches.

The restaurant goes by the name "Napa", synonymous with great quality and laid back style. An aesthetic of cool marble, ebony laminates and crisp green leathers anticipate the fresh seasonal menu. A striking feature is a series of screens of polished stainless steel and rotating green Plexiglas ellipses affording glimpses into the resident's bar, a tucked away corner of moody smoked mirrors and cow hide upholstery.

Bedrooms are coloured according to the themes of surf, turf and desert. All have a full wall of glazing dressed with colour washed voiles and graphic black and white patterned curtains.

Carpets are shag pile, Bespoke furniture is a cool and rectilinear combination of faux ebony veneer and mirror. Bathrooms are characterised by profiled ceramic tiles to an original 1960s design in tangerine, lime and slate grey. The fittings are similarly unconventional; the basin and WC are triangular.

1. Entrance of the original office building
2. Working area of the original office building
3. Corridor of the original office building
4. Refurbished frontage
1. 原有建筑入口
2. 原有办公区
3. 原有内部走廊
4. 改造后酒店正面

5

背景

克里斯莫兰酒店有120间客房，位于伦敦东南部的克里斯高路旁一个活力十足的街区内。原建筑建于20世纪60年代，起初被用作办公楼。

目标

酒店的设计理念是以"西伦敦及西海岸"为主题，对20世纪60年代的加州建筑风格加以现代的诠释，同时将传统的伦敦和辉煌的加州特色相结合。

过程

热带棕榈和悬挑的混凝土雨篷下快捷的车辆通道构成了酒店的特色。深色木纹大理石和涂漆橡木板将大堂、酒吧和餐厅空间联系在一起。双层高的大堂显得格外庄严，由抛光钢和有机玻璃材料定做的枝型吊灯从天花上悬垂下来。"环球酒吧"这一名字源于悬挂在接待处和酒吧之间那个闪亮着并不断旋转的大球。吧台长达15米，采用大理石铺面，周围摆放着橘红色的皮沙发座椅。

餐厅被命名为"纳巴"（美国加利福尼亚州西部一城市，位于奥克兰以北，是纳巴山谷的中心，此山区是有名的葡萄园地区），象征着高品质和轻松的氛围。大理石板、仿乌木的层压板和亮绿色的皮革材料寓意着餐厅将提供四季的新鲜食

品。更为吸引眼球的是一面由多个椭圆形磨光不锈钢片和旋转的绿色有机玻璃片组合的屏风。透过屏风可以瞥见酒吧座位区，布置了镜子和牛皮座椅。

客房内的色彩与设计主题相互呼应，如冲浪主题、草地和沙漠主题。所有客房安装着落地窗，并用巴里纱（一种轻而透明的薄纱）窗帘和黑、白图案的窗帘点缀。地面上铺着长绒地毯，家具极为简洁——仿乌木饰面板的长方桌和一面镜子。卫生间采用了20世纪60年代设计的橘红色、绿色、石灰色的铸型瓷砖贴面。这里的设施也与众不同，手盆和座便都是三角形的。

6

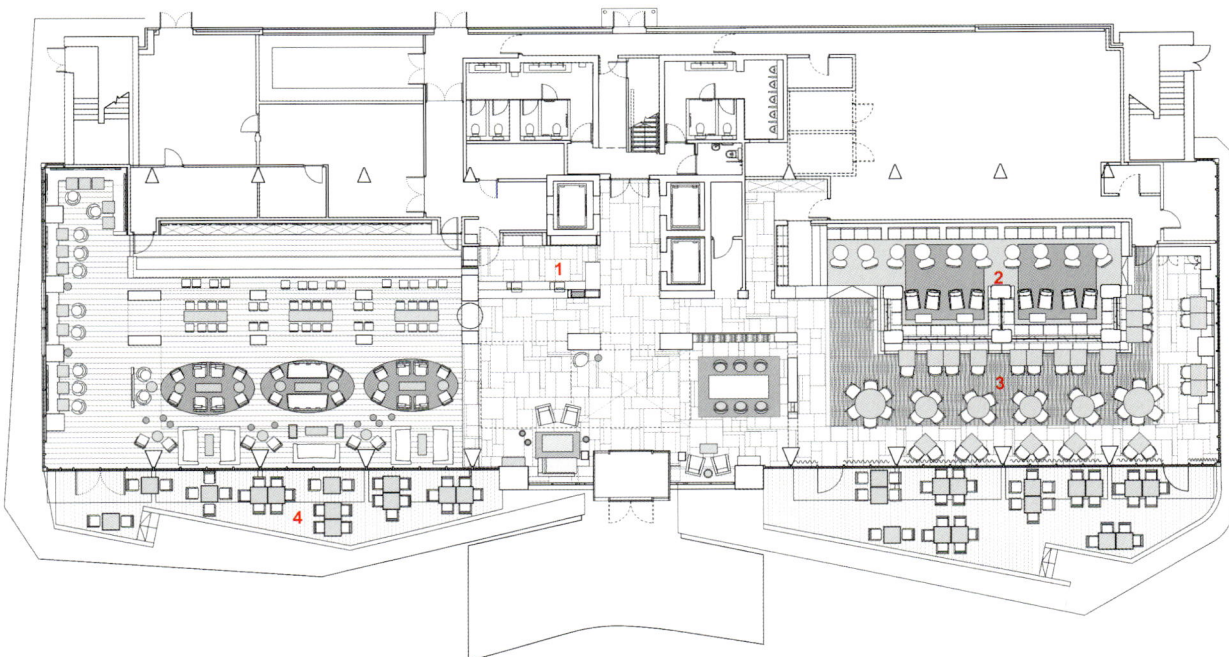

1. Reception
2. Resident's bar/informal dining
3. Main restaurant
4. Terrace
1. 接待区
2. 酒吧/休闲就餐区
3. 主餐厅
4. 露台

9-10. Restaurant
9-10. 餐厅

UNA Hotel

UNA 酒店

Location: Naples, Italy
Completion: 2006
Designer: Luca Scacchetti Architects
Photographer: Luca Scacchetti Architects

地点：意大利 那不勒斯
翻新时间：2006年
设计：Luca Scacchetti 设计公司
摄影：Luca Scacchetti 设计公司

BACKGROUND

A 19th century historic building, facing on Garibaldi Square in the city centre, has been restored and transformed into a 4-star hotel for a prominent Italian hotel chain. The original building, a 7-storey tuff-stone structure, very long and narrow, has suggested the design of a vertical hall crossed by staircases and elevators dangling in space.

1

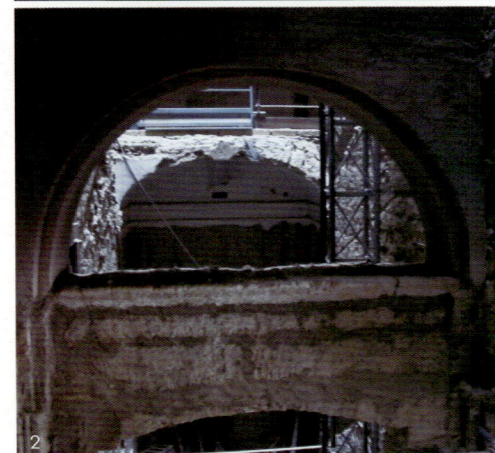

2

OBJECTIVE

Attention to every detail, the interior decoration, and respect of the original architecture, restoration of local materials has been given to the renovation project of the Una Hotel Napoli.

SOLUTIONS

The building has been a careful plan of strengthening. The need to maintain the original structure was built a reinforcement of most of the load-bearing walls with structural plaster applied on most surfaces, while the floors were reinforced with a lightweight concrete slab in the maintenance of the original shares. During the work the unveiling of the walls in tuff suggested to work and enhance the contrast between these rough surfaces and the original stucco decorations of nineteenth-century taste that had been stored in the stairwell.

In terms of the distribution the original plant allow a double staircase overlooking a tall and thin hall covered with a skylight overhead. The cut made up the original hall that crossed the building was restored and became the principal area of the hotel, as a result of the preexistent presence of business on the ground floor was only possible to obtain a hall of small dimensions and a small living space. This common vertical space becomes a privileged place to overlook, as the revival of a Neapolitan alley; on the tuff walls opened windows from opened embrasure splayed projecting which interacts with the wrought iron railings of the landing of the main stair.

In the project, great attention has been given to the windows of this area - the fire prevention should have security features REI 60, but still had to ensure transparency and lightness of all elements.

The lift is contained in a structural glass facade that has been carefully designed for maximum transparency from the inside and at the same time the almost total invisibility of the metal structure. The design of this element, the main point of focus has been the choice of glass, standard reflecting but placed with the reflected part inward, enough to hide the technical parts at sight, but let see the cab in motion.

The bow window jutting in the space were made in steel frame and aluminium casing folded, leaving a detachment from the walls to allow light brushing of the tuff walls set in sight.

On the sixth floor is located a restaurant open also to outside guests, characterised by the angle of the pitched roof, to crown the building, above the terrace offering a panoramic view of Mount Vesuvius.

The facades of the building have been the subject of careful renovation, through the restoration of the original colours of the plaster, wooden windows and iron railings, and the regular drawing of the facade has been highlighted by a nightlights design.

This hotel consists of 89 rooms arranged along the two corridors that overlook the central area; the rooms are made in accordance with the original structure and therefore the size of the basic module is conditioned by the depth of the existing buildings. The interior walls are partly painted in strong colours, such as the Pompeian red and sage green and in some place leaves and re-emerge the original masonry tuff to point out a successful contrast between the old and new.

1. Hall of the original building
2. Wall structure of the original building
3. Main entrance
1. 原有建筑大厅
2. 原有建筑墙结构
3. 酒店主入口

背景

UNA是意大利知名连锁四星酒店，位于那不勒斯。这家酒店由坐落在市中心的加里波第广场（Garibaldi Square）对面一幢建于19世纪的历史建筑改建而成。原有建筑共为7层，以凝灰岩结构为主，呈细长造型，这决定了纵向元素的增建。

目标

注重每一处细节、尊重建筑原有结构、采用当地材质构成了整体设计目标。

过程

设计首先对原有建筑进行加固，保留现有结构的同时打造了一系列的承重墙壁，表面采用灰泥粉饰。地面铺设了轻型水泥板，保护原来的结构。之后将凝灰岩墙面装饰去掉，旨在突出墙体粗造的表面与楼梯井内保留下来的19世纪灰泥装饰的对比。

根据原有的格局，设计师决定新建一个剪刀楼梯，站在顶端可俯瞰下面高大细长并带有天窗的大厅，而这里也成为了酒店中最为重要的区域。大厅犹如那不勒斯特有的小巷一般，在这里可以欣赏下面的景象，凝灰岩墙壁上的窗户与楼梯缓步台处的铁栏杆相互呼应。大厅区域内，设计师在窗户的设计上花费了很多的心思，符合消防标准的同时确保所有元素的透明感与轻巧性。

电梯设置在玻璃外观结构内，从而最大限度保证其透明度。这一设计中最主要的部分即为玻璃的选择，达到标准反光标准并将使反光一面朝向内侧，便于将技术设备隐藏。

凸窗采用钢筋框架及弯曲的铝壳打造，与墙壁之间留有一定的空隙，以便于光线的射入。餐厅位于顶层，对所有人开放，角状的斜屋顶别具特色。露台上可欣赏维苏威火山的壮丽景致。

建筑外观经过仔细翻修，原有的色彩、木窗、铁栏杆以及表面的图案在夜晚灯光的照耀下更加引人注目。

经过改建之后的酒店共包括89间客房，分布在走廊的两侧，可以俯瞰中央大厅。客房根据原有格局打造，因此空间的大小受到了一定限制。客房内的墙壁上一部分被刷上了浓郁的色彩，如庞培红（从偏灰到中度的红色）及灰绿色，一部分则将凝灰岩材质直接裸露在外，成功实现了新与旧的对比。

4. Entrance hall
5. Lobby reception
6. Lobby lounge
4. 入口大厅
5. 大堂接待台
6. 大堂休息区

7. Newly designed double staircase
8. Corridor
9. Staircase
7. 新建"剪刀楼梯"
8. 走廊
9. 楼梯

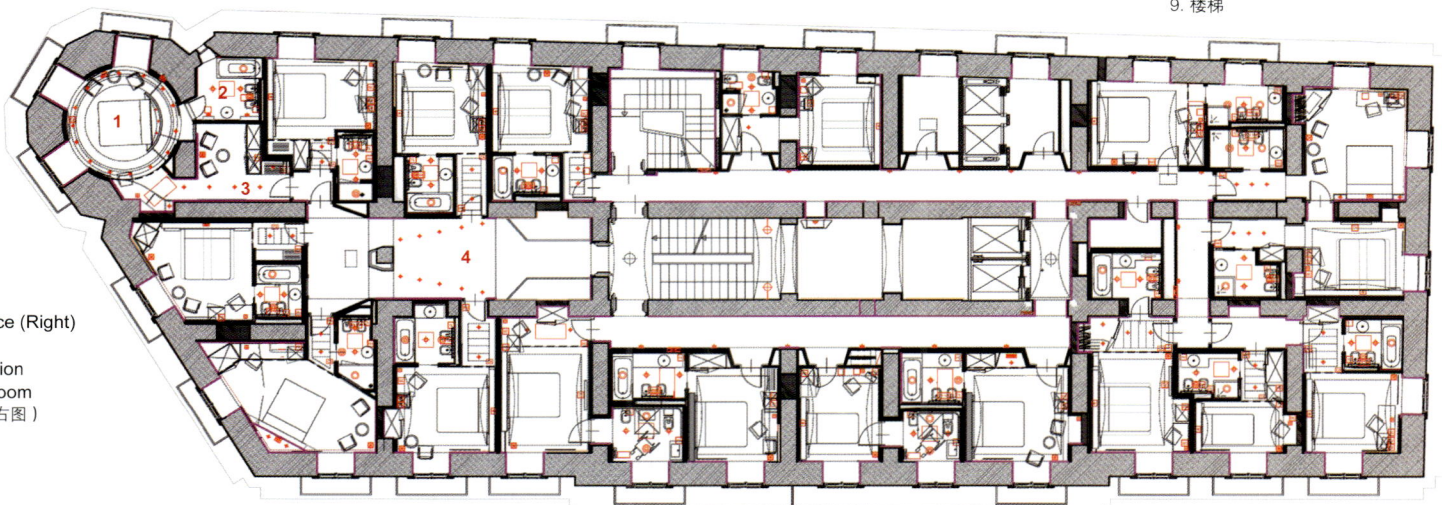

1. Entrance (Right)
2. Foyer
3. Reception
4. Guestroom
1. 入口（右图）
2. 门厅
3. 接待区
4. 客房

11

10–11. Corridor
10–11. 走廊

12 Restaurant
13."Green" guestroom offering a relaxing and welcoming atmosphere
14. Two glass doors bringing outside view into the guestroom
12. 餐厅
13. "绿色主题" 客房为客人提供轻松友好的氛围
14. 客房内，两扇玻璃门将室外的美丽景致 "吸纳" 进来

13

14

Bei Hotel

北湖宾馆 "天湖楼" 改造

Location: Xinyu, Jiangxi, China
Completion: October 2006
Design Company: LISPACE
Design Director: Jia Li
Photographer: Gao Han
Area: 11,340m²
地点：中国 江西省 新余
翻新时间：2006年10月
设计公司：北京立和空间设计事务所
设计主持：贾立
摄影：高寒
面积：11340平方米

BACKGROUND

Bei Hotel is a government hotel, located in Xinyu, Jiangxi province, China. This hotel is well-known for its beautiful natural environment and location. With the development of economy, more hotels were found. Bei Hotel faced the pressure of market competition, so it had to be renovated.

OBJECTIVE

Required by the local government, Bei Hotel would be charged with the management and operation of new Conference & Exhibition Centre. After the communication between the design team and the client, Bei Hotel is defined as a conference and business hotel in Xinyu. The hotel should increase 128 guestrooms.

The beauty of bamboo in Jiangxi province is usually written by poets in ancient and present. People all over the world are intoxicated by bamboo's beauty especially after Ang Lee's film *Crouching Tiger, Hidden Dragon*. Bamboo and its staggered shadow become the element of this project.

SOLUTIONS

Architecture Design:
The main purpose of renovation is to increase the guestrooms. Following the principle of keeping the original height of the old building and surrounding environment, designers decided to build a new building in the north of the old building.

Staggered steel frame structure and stained glass's triangle form on black glass curtain of façade, presenting the modern meaning of bamboo and its shadow.

The void space between the new and old building becomes a sunny atrium. It is a good way to share sunshine with corridor. The bridge linking new and old building weakens the height of atrium and increases the level of space.

Interior Design:
The design of new lobby blends the local architecture style: partition wall made of Chinese traditional grey tile and metal, and customer-made crystal screens of bamboo leaves shape. Metal and glass reception desk is shaped by bamboo shoot. Those elements give the hotel the spirit of Jiangnan's style.

Coffee bar in the first floor is formerly the guest room of the old building, which had beams that couldn't move. Designer takes advantage of beams to transfer a light platform. Through the glass floor people could see the undecorated beam and white pebble, which continue the life of the original building. This is greeting for architecture.

Interior design of atrium keeps the former structure of the old building for the new building. Customer-made glass and tree lamps make the space full of mottled shadow. The installation art of "Disappearing Green" made of steel frames is painted to green and wood-colour. The colours will be changing when people walk through the atrium. It tells story of Green.

The interior design of new guestrooms breaks traditional layout and furniture design. Glass wall brings nature light into washroom in daytime and increases space in vision. Also guests could watch TV when they want. On furniture design, drawers have been deleted, which not only reduced the odds of lost but also easily cleaned. Curtains replace partition wall in suit room. It gives multi-function to the living room and the bedroom. All of those save energy and cost.

Renovation of Bei Hotel is a very successful performance of using Chinese traditional element to express modern design.

1. General view of the original hotel
2. Exterior view after expansion
1. 酒店改造前全景
2. 酒店扩建后外观

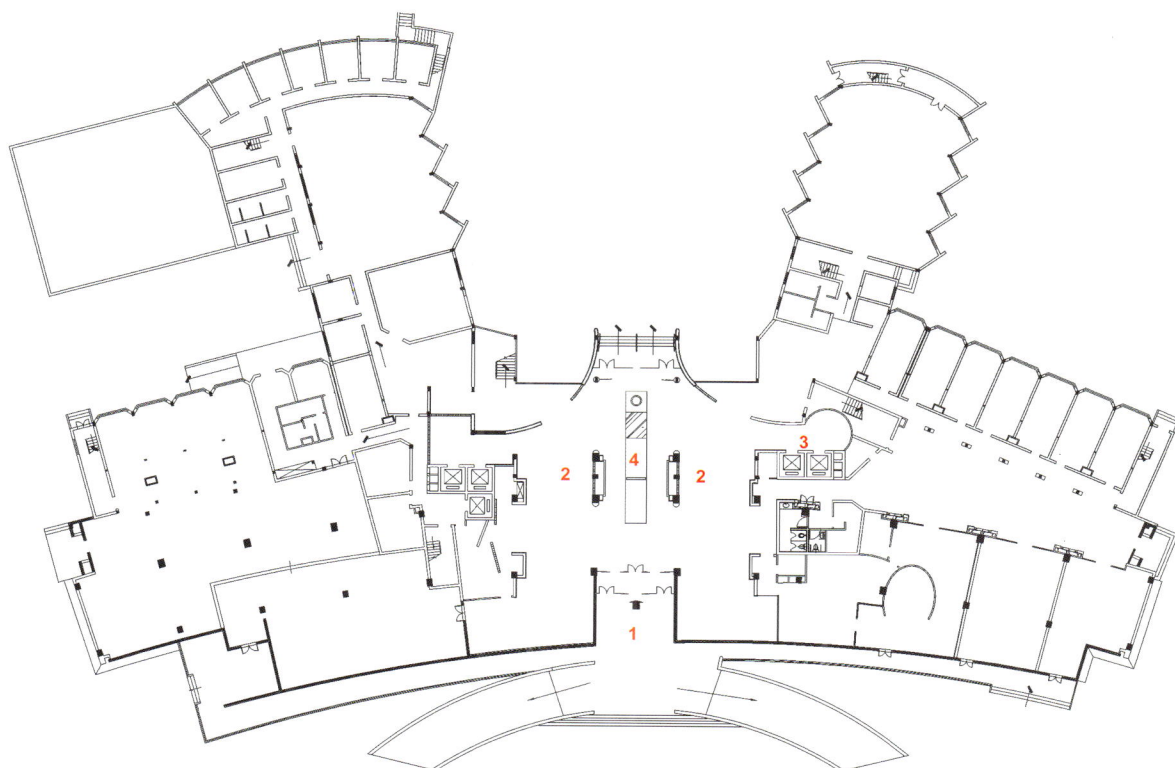

1. Main entrance
2. Lobby
3. Bathroom
4. Atrium
1. 主入口
2. 大堂
3. 卫生间
4. 中庭

3

背景

位于江西省新余市的北湖宾馆是个邻湖的市政酒店。凭借优越的地理位置和美丽的自然环境一直享有盛誉。但随着当地经济的发展和其他酒店在新余市内的出现,北湖宾馆面临的市场竞争压力也应运而生,改造迫在眉睫。

目标

因为市政的需要,北湖宾馆承担了新余市新会展中心管理和运营的工作。设计团队在和酒店方认真沟通后,决定将酒店的市场及形象定位为商务会议型酒店,客房数量需增加128间。

江西竹林之美是古今文人乐提的。李安的电影又使全世界的人陶醉并寻找这种美。竹与竹的交错,恍惚斑驳的光影成为本案建筑及室内设计的主要元素。

过程

建筑设计:

主楼改造工程的主要目的是增加客房量,但是为了不破坏周边的自然环境,在不增加主楼高度的原则上,增建其南侧新的客房楼。一层群楼用交错的钢架结构和玻璃写意

"月光穿竹翠玲珑"的意境。黑色玻璃幕墙上凹进的三角空间是对竹叶的现代诠释。

利用新老楼之间的空隙创建了共享空间式的中庭,使客房走廊能享用阳光,避免昏暗压抑。连接新老楼的天桥柔化了中厅内的高耸感,并增加了空间的层次。

室内设计:

大堂的设计融入了当地民居的建筑风格:灰瓦和金属材质的隔墙、竹叶造型的水晶幕帘、形似竹笋的金属前台,使酒店空间充溢着现代的徽派气质。

大堂二层的咖啡厅曾是老楼的客房空间,拆除原有隔墙后露出的反梁,顺势成就了发光地台。透过玻璃地面,能够清楚的看到不经装饰的反梁,这是对建筑生命的尊重和致敬。铺满地面的白色石子上,生命波纹又一次的和老结构交融、重叠,延续给予了老楼新的生命。

中庭的室内设计保留了新老建筑给予共享空间的结构美感。手工制作的玻璃 "树灯",让整个空间充满绿色斑驳的

"树"影。内外分别是绿色和木色的槽钢拼装成的现代艺术品"消失的绿色"成为中庭与电梯之间的隔断,两种色彩随着客人视角的移动交替变化着,诉说着绿色对我们的意义。

新增的客房打破了传统的平面布局和家具设计。玻璃隔墙使卫生间白天不再黑暗,在视觉上增大了空间,并且客人可以在任何地方观看电视。舍弃抽屉的家具设计减少了客人遗忘物品的几率,并避免了清洁房间时的死角。套房里布帘的运用,让客人分隔使用卧室与客厅更加方便自如。这些最终使酒店在节省能源的同时达到了降低客房成本的目的。

北湖宾馆"天湖楼"的改造,是一次运用中国传统元素进行现代设计的成功尝试。

3. Refurbished lobby
4. Distinctive lighting structure
5-6. Partition wall made of Chinese traditional grey tile
   and metal in the lobby
3. 改造后大堂
4. 独特的灯饰
5-6. 大堂内,灰瓦及金属材质隔断墙

7. Café
8. Living room in the deluxe suite
7. 咖啡厅
8. 套房内客厅

Hotel Restaurant "AL Ronchetto"

阿尔·兰彻托酒店及餐厅

Location: Salgareda, Italy
Completion: 2005
Designer: Arch. Renata Giacomini
Photographer: Gabriele Gomiero
地点：意大利 萨尔加里达
翻新时间：2005年
设计：雷纳塔·贾科米尼建筑师事务所
摄影：加布里埃莱·格米尔罗

OBJECTIVE

The expansion project of an old country house located in the buffer zone of the bank of the Piave Salgareda, was created by architects Franca Furlan and Franco Lorenzon. The intervention of interior design has been requested by the client to give a picture of coordination at a construction site that has grown year after year. The will of the client was to produce a refined but simple at the same time, where the flavour of tradition and characterisation of the surrounding area were dominant.

The old house had enough space to contain the required spaces and features that clients have subsequently identified for the first realisation of a small restaurant, then a small hotel of 20 rooms and finally a banquet hall with 300 seats

and its kitchen technique. The intervention of interior design is required when the site was still in a phase of distribution choices to make. Thus, it is necessary to give a general idea of coordination between the architectural skin of the building and furnishings.

SOLUTIONS

The recovery values is accomplished through the development of the following topics:
• Naturalness
• Relationship with the history and the past
• Searching for authentic flavours:
  food quality
• Search for elegance and refinement

The materials that have covered the house and extension have been defined in relation to the spaces. Natural stones (Santa Fiora, sandstone, gneiss) were used for internal routes, large corridors that look like streets outside, to make the relationship ambiguous between interior and exterior. The terrazzo was selectd for the large banquet room that re-configured in a semicircle of noble Venetian palaces. Still with reference to natural stone tiles are used as coatings Tate in the bathroom.

Coping stones split and polished, reveal a "natural" feel and have been used to all floors, from reception to the bar. The furnishings are minimalist, clean and strict geometries essential. All the elements were treated iron rust finish to remember the old iron work in the countryside.

The integration with the architectural design has been implemented in the exaltation of form by a proper use of colours and lighting to highlight the major structural elements. The red brick for the old house emphasised the importance of its size. It is a colour typical of the great houses of the campaign. As for the part with the big

bows a light colour was selected. The new buildings, silos, were stained with a light grey to emphasise the similarity.

The choice of lighting is invisible. Among the wooden beams of the first floor of rust-coloured cylinders Viabizzuno, light can't be seen. Fluorescent light hidden by veils of the corridor can't be seen as well.

The 10 rooms of the house "old" have been designed with a more traditional sign - the bathrooms have mirrors framed in lacquered wood pulp, as in country houses. While in the rooms of the house "new" are designed for minimal large canopies and bathrooms, while maintaining the sink bowl design has a much more rigorous than the former.

The reception is made of wood with a great plan in Santa Fiora. In the bar area the designer has designed a large fireplace with a grand plan of Santa Fiora and the wood in sight. Two large sofas allow guests to enjoy the warmth of the fire. Particular attention was paid to the great dividing doors which were made as they were a combination of different woods, almost simulating a collection of waste storage.

1. Exterior view of the original building
2. Ceiling of the original building
3. General view of the hotel after expansion
1. 原有建筑外观
2. 原有建筑天花
3. 扩建之后酒店全景

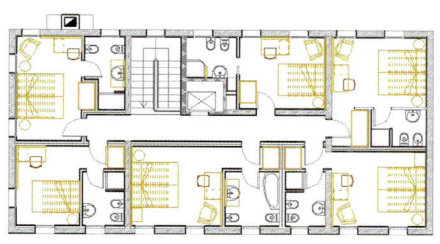

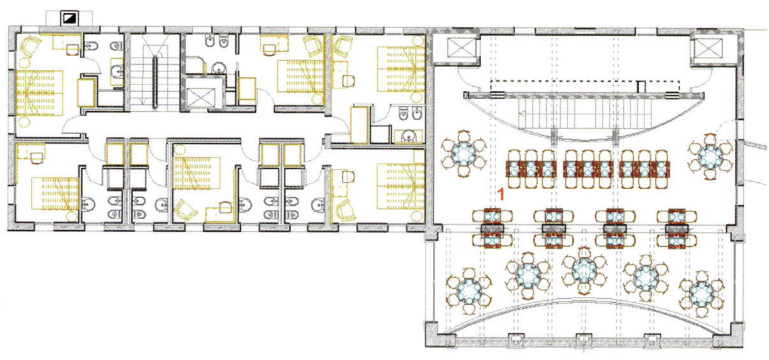

1. Restaurant
2. Guestroom
1.餐厅
2.客房

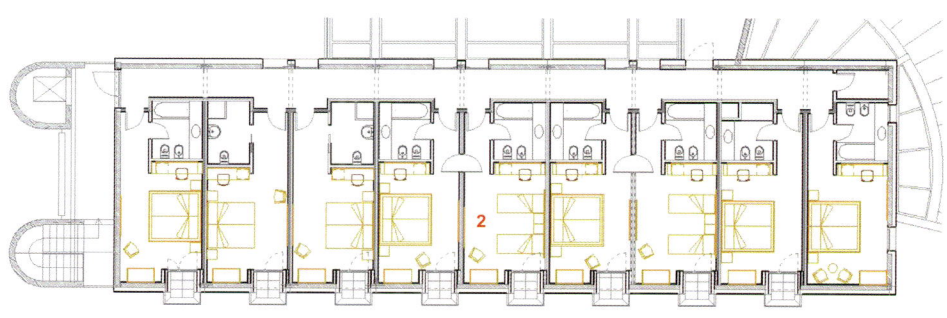

4

5

目标

酒店原有建筑为一幢古老的乡村住宅，位于皮亚韦河沿岸的缓冲地带。根据各户要求，室内设计的主要目标是打造雅致简约的环境，同时充分展现这一地区的传统与特色。

原有住宅具有足够的空间，可容纳一个小型餐厅、一个拥有20间客房的小酒店以及一个可容纳300人的宴会厅。室内改造方案被决定之后，建筑还在翻修之中，因此，如何平衡建筑外观同室内装饰显得尤为必要。

过程

建筑翻新的价值通过如下几项体现：
——自然特色
——同历史及传统的联系
——寻求独特的地域特色（食物）
——力争典雅与精致

原有建筑及扩建部分均选用天然及当地材质装饰。天然石材用于铺设室内通道、走廊等区域，借以模糊室内外空间的界限。宴会厅呈现半圆造型，犹如维也纳宫殿一般，主要采用磨石子材料装饰。

压顶石经切割及抛光之后被大量运用，从接待台到酒吧区，营造出自然气息。装饰家居以简约风格为主，清晰的造型别具特色。经铁锈饰面的结构让人不禁联想到乡村古老的铸铁工艺。

颜色及灯饰的选择在一定程度上能够对建筑结构起到修饰作用。原有建筑大部分采用红色砖墙饰面，突显出当地的特色。部分则采用浅色装饰，扩建部分选择淡灰色饰面，与老建筑寻求一丝相似感。

灯饰的选择突出"不可见性"。一层，灯饰隐藏在木梁之间，若隐若现。同样，走廊内，荧光灯被"遮盖"起来。老建筑内的客房设计突出传统特色，浴室的镜子镶嵌在木框内，犹如传统乡村住宅一般。扩建部分的房间内则以现代风格为主，活力十足。

接待台采用木材打造，酒吧区内安装有壁炉，坐在宽敞的沙发上可以感受到火光的温暖。最为引人注目的当属由不同木材拼接而成的隔断门，格外吸引眼球。

4. View into the lobby
5. Corner view in the lobby
6. Reception
7. Restaurant
8-9. Local wood being extensively used in the guestroom
4. 大堂
5. 大堂一角
6. 接待台
7. 餐厅
8-9. 客房内大量运用当地木材装饰

7

Index 索引